Phase Change Materials

Related Titles

Biopolymers Based Advanced Materials
ISBN: 978-0-6482205-4-1 (e-book)
ISBN: 978-0-6482205-5-8 (hardcover)

Functional Polymer Blends and Nanocomposites
ISBN: 978-0-6482205-6-5 (e-book)
ISBN: 978-0-6482205-7-2 (hardcover)

Functional Nanomaterials and Nanotechnologies: Applications for Energy & Environment
ISBN: 978-0-6482205-2-7 (e-book)
ISBN: 978-0-6482205-3-4 (softcover)

Advances in Polymer Technology: Material Development, Properties and Performance Evaluation
ISBN: 978-1-925823-00-4 (e-book)
ISBN: 978-1-925823-01-1 (hardcover)

Polymer Nanomaterials for Specialty Applications
ISBN: 978-1-925823-03-5 (e-book)
ISBN: 978-1-925823-04-2 (hardcover)

Advanced Materials
ISBN: 978-1-925823-05-9 (e-book)
ISBN: 978-1-925823-06-6 (hardcover)

Biofuels
ISBN: 978-1-925823-12-7 (e-book)
ISBN: 978-1-925823-13-4 (hardcover)

Liquid Crystalline Polymers
ISBN: 978-1-925823-16-5 (e-book)
ISBN: 978-1-925823-17-2 (hardcover)

Polymer Nanocomposites: Emerging Applications
ISBN: 978-1-925823-14-1 (e-book)
ISBN: 978-1-925823-15-8 (hardcover)

Phase Change Materials

Dr. Vikas Mittal
Editor

CWP
Central West Publishing

This edition has been published by Central West Publishing, Australia

For more information about the books published by Central West Publishing, please visit https://centralwestpublishing.com

Disclaimer
Every effort has been made by the publisher, editor and authors while preparing this book, however, no warranties are made regarding the accuracy and completeness of the content. The publisher, editor and authors disclaim without any limitation all warranties as well as any implied warranties about sales, along with fitness of the content for a particular purpose. Citation of any website and other information sources does not mean any endorsement from the publisher and authors. For ascertaining the suitability of the contents contained herein for a particular lab or commercial use, consultation with the subject expert is needed. In addition, while using the information and methods contained herein, the practitioners and researchers need to be mindful for their own safety, along with the safety of others, including the professional parties and premises for whom they have professional responsibility. To the fullest extent of law, the publisher, editor and authors are not liable in all circumstances (special, incidental, and consequential) for any injury and/or damage to persons and property, along with any potential loss of profit and other commercial damages due to the use of any methods, products, guidelines, procedures contained in the material herein.

NATIONAL LIBRARY OF AUSTRALIA

A catalogue record for this book is available from the National Library of Australia

ISBN (print): 978-1-925823-39-4
ISBN (e-book): 978-1-925823-38-7

Contents

Preface

Phase change materials (PCMs) can store and release heat, while they are changing phase. They allow relatively large amount of energy to be stored in comparatively small volumes. Due to their attractive and controllable properties, they are useful in many applications. For instance, PCMs have been used as smart building materials, energy storage devices, heat recovering agents, etc. By virtue of their properties, these materials have also found applications in smart textiles and aerospace industry. This book aims to review the synthesis and properties of a variety of functional organic and inorganic phase change material systems developed in the recent years for advanced applications.

Chapter 1 reports that a number of methods have been employed to synthesize composite assemblies incorporating PCMs, however, economical, environmental friendly and bulk production methods need to be explored further. Thermal and chemical stability as well as conductivity of PCMs also need to be optimized for their practical use. Chapter 2 concludes that the use of PCMs in solar energy technologies has allowed the option of storing thermal energy produced by the solar collectors during the day so that it is possible to have energy available during the night. In addition, the use of PCMs also allows increasing the efficiency of the photovoltaic modules by increasing the electricity generation. Chapter 3 mainly summarizes the effect of outdoor temperature on composite phase change materials (CPCMs) for reducing the influence on indoor environment as well as the combination of CPCMs and building materials to form shape stabilized composite phase change material wallboards (SSCPCMWs). In Chapter 4, the thermodynamic analysis of different photovoltaic thermal (PVT) systems is presented from the energy and exergy viewpoints. In order to investigate the performance of PVT systems, different approaches have been studied and appropriate equations are expressed. Moreover, in this chapter, the effect of using nanofluids (as the coolant fluid) as well as PCMs on the performance of PVT systems is evaluated. Chapter 5 further reviews the properties and applications of a wide variety of inorganic PCMs. Chapter 6 presents attapulgite based composite phase change materials for solar thermal energy storage. The produced form stable composite PCM showed good thermal and chemical reliability after 500 times repeated-thermal cycling tests.

The book would not have been successfully accomplished without the support of chapter contributors. The book is dedicated to my family for unswerving support and constant motivation.

Vikas MITTAL

1

Thermally Phase Changing Materials: Synthesis, Properties and Applications

Liyamol Jacob and Vikas Mittal*,**

Department of Chemical Engineering, The Petroleum Institute (part of Khalifa University of Science and Technology), Abu Dhabi, UAE

**Corresponding author*: vik.mittal@gmail.com
***Current address: Bletchington, Wellington County, Australia*

1.1 Introduction

Materials that undergo phase change at particular temperature are defined as thermally phase changing materials (PCMs). These materials can store and release heat in the form of energy, as required by specific applications. This is facilitated by their phase change owing to the change in temperature. They allow relatively large amount of energy to be stored in comparatively small volumes. Due to their attractive and controllable properties, they are useful in many applications. For instance, PCMs have been used as smart building materials, energy storage devices, heat recovering agents, etc. By virtue of their versatile and moldable properties, these materials have also found applications in smart textiles and aerospace industry [1]. Their attractive characteristic to maintain the temperature at a constant level during the phase change is of particular relevance.

Materials undergo different types of phase changes, viz. solid-solid, solid-liquid and liquid-gas. Solid-solid and solid-liquid PCMs have garnered widespread attention due to superior textural and thermal properties. In addition, materials with phase change properties in between solid-solid and solid-liquid phase changes are known as shape stabilized or form stable PCMs. They comprise of supporting materials and working substances. They can physically stay in solid state even if the working materials undergo phase change from solid to liquid, which makes it very easy to use them in many practical applications [2].

Phase Change Materials, edited by Vikas Mittal

Various methods are utilized for the synthesis of PCM materials. Blending, adsorption, impregnation, sol-gel method and copolymerization are some of the widely used methods for this purpose. These methods are effective in producing PCMs with different physical, thermal and textural properties. Additionally, these methods help to tune the structure, properties and morphological attributes of the PCMs to make them suitable for particular applications. Thus, the ability of the phase change materials to be molded according to the intended use makes them one of the most advanced and efficient materials [3].

1.2 Different Phase Change Materials (PCMs)

1.2.1 Nanomaterial PCMs

Melting temperature, thermal conductivity and energy storage density are the main aspects that decide the thermal performance of a PCM. The enhancement in the properties of the PCMs by composite formation leads to the expansion of their spectrum of applications [4]. Specifically, low thermal conductivity of the phase change materials can be manipulated by incorporating them with different kinds of nanomaterials. Nanomaterials such as nanoparticles, nanofibers, nanotubes and nanosheets have gained significant research attention for this purpose [5-8]. Many studies have also reported nano-encapsulated PCM systems.

Vanadium dioxide encapsulated with gold nanoparticles shows excellent surface plasmon resonance combined with phase change properties. Of specific intertest are its quantum phase changing properties, which interact effectively with the electrical and optical properties of the material. Gold nanoparticles in the nanocomposite structure act like photocathodes, which cause very fast phase transition in vanadium dioxide nanostructure, when illuminated by a spectrally resonant femtosecond laser pulse [9].

Paraffin wax incorporated with carbon nanotubes (CNTs) is a multifunctional phase change composite where energy storage can be driven by small voltages [10]. The CNTs act as flexible encapsulation scaffold for the wax and also retain a highly conducting network over it during the phase change process. The phase change enthalpy and thermal conductivity of the paraffin wax is improved due to the uniform interpenetration of the nanotube network and paraffin wax with high affinity towards each other. The developed material can be effectively utilized in sunlight absorption and conventional lithium ion batteries [10].

N-octadecane nanocapsules as organic PCMs have been used for thermal energy storage [11]. These nanocapsules were developed by trapping n-octadecane as a core with St (styrene)-MMA (methylmethacrylate) copolymer shell using miniemulsion *in-situ* polymerization method. Nanocapsules were observed to have a melting point of 29.5 °C and a crystallization temperature of 24.6 °C, measured by differential scanning calorimetry (DSC). The enthalpy of melting and crystallization of the PCMs was found to be 107.9 and 104.9 $J.g^{-1}$ respectively.

Other studies on different polymer nanocomposite PCMs have also confirmed their potential of application in various fields. Sol-gel method was used to form paraffin and TiO_2 composite, whereas *in-situ* polymerization was reported to form urea-formaldehyde for nano-encapsulation of n-tetradecane [12,13]. Some of the PCMs can also unavoidably form formaldehyde which can pose threat to the environment and animal life [14].

As an example, Figure 1.1 shows the transmission electron micrographs (TEM) and high-resolution TEM (HRTEM) of phase changing ZnS nanostructures [15].

1.2.2 Carbon PCMs

Carbon materials have been employed in many studies to enhance the property profiles of different PCMs. Activated carbon, CNTs and graphene have been extensively used in the synthesis of advanced PCMs [16].

To achieve stable phase change composites, polyethylene glycol (PEG) was coupled with activated carbon and silica [17]. A prominent effect on the crystallinity of PEG by the use of porous carbon mesh and silica sieves was observed. The material showed improved latent heat, low melting point, low supercooling and high heat storage efficiency.

Phase changing materials with carbon nanotubes have obtained special interest owing to the potential for many applications. Composite systems formed by incorporating single walled CNTs (SWCNTs), multi-walled CNTs (MWCNTs) and carbon nanofibers have been used to analyze the improvement in the latent heat energy storage systems using PCMs [18]. The DSC studies revealed a remarkable improvement in the latent heat as compared to pure wax. Among the different

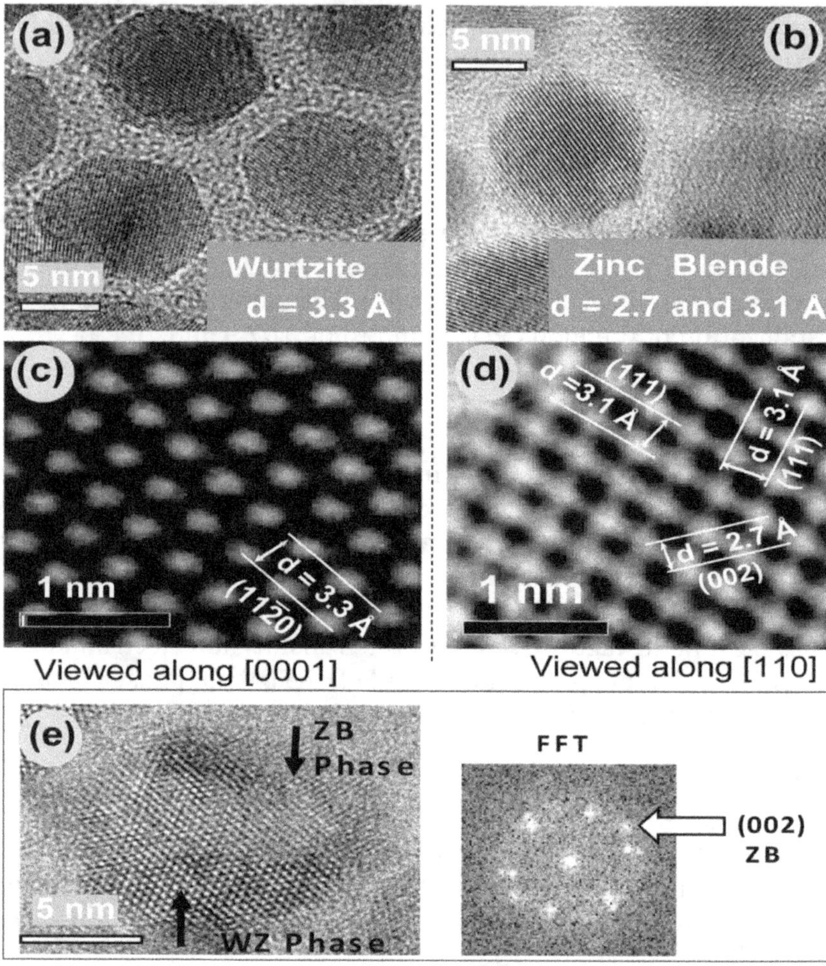

Figure 1.1 HRTEM images of (a) WZ and (b) ZB ZnS platelets; (c,d) the corresponding magnified HRTEM images. (e) An intermediate TEM image showing the existence of both WZ and ZB phases in a single platelet. Right, the FFT with identification of the (002) plane of ZB ZnS. Reproduced from Reference 15 with permission from American Chemical Society.

composites, the SWCNTs showed the best improvement of 13% for 1% loading owing to higher surface area and molecular density. The study was combined with a theoretical model which was found to be in good agreement with the experimental studies. Paraffin wax modified with CNTs showed shift in the melting point to a lower temperature [19]. This was further enhanced by increasing the carbon nanotubes content. The addition of CNTs reduced the latent heat capacity as well. The thermal conductivity was observed to improve in the case of both liquid and solid phases.

Elgalfy and Lafdi [20] studied the behavior of paraffin wax with carbon nanofibers. The cooling rate was predicted theoretically by analyzing the changing thermal response of the composite during solidification process. The cooling rate was observed to show improvement owing to the change in the thermal conductivity with regard to the pure paraffin wax. Based on these results, the authors introduced an analytical model which effectively predicted the thermal conductivity. The model could be further modified to analyze and predict the effect of surface modification of carbon nanofillers on phase change materials.

SWNTs/poly(methylmethacrylate) composite has been reported for studying the improvement in the thermal properties of the polymer by adding different percentage of SWCNTs [21]. Both thermal and electrical conductivity improved by the introduction of SWCNTs to the polymer matrix. For instance, improvements of 57% in the thermal conductivity and several magnitudes in the electrical conductivity were observed by incorporating 7 wt% SWCNTs in the polymer matrix. The percolation of SWCNTs in the polymer network was observed to have influenced the thermal properties of the polymer matrix manifold.

Another study used the ability of carbon nanotubes to confine the materials inside the tubular structure to create materials that undergo dimensionally confined phase changes [22]. Inside the carbon nanotubes, the material can show a solid-liquid critical point beyond the point where solid-liquid distinction disappears, unlike its bulk counterpart. This feature can be used for the direct and continuous transformation from sloid to liquid phases. The study developed numerical simulations where the ice transitions showed different phases which could not be found in the bulk transition, along with different critical temperature points. Water, inside CNTs with 1.1 nm to 1.4 nm dimensions, was shown to undergo phase change from hexagonal cubic phase to heptagonal cubic phase, and a continuous phase change was observed from solid-like square or pentagonal ice nanotubes.

Nano-architectured phase change material composite made of graphene and 1-octadeconal was reported to demonstrate improved thermal conductivity without much change in the heat of fusion. Figure 1.2 shows the improvement in the conductivity of the composite as a function of graphene content [23].

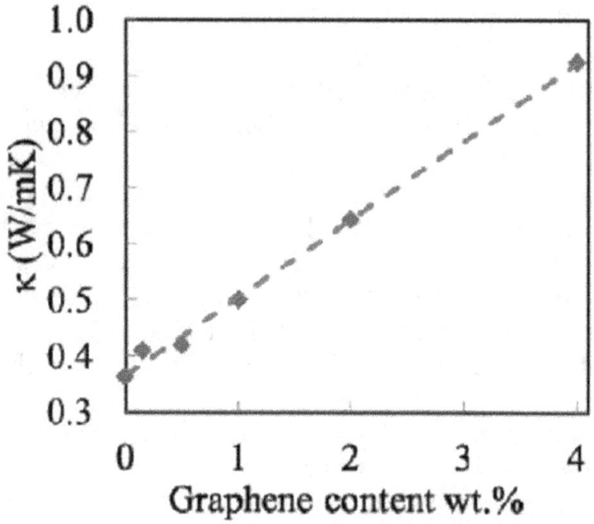

Figure 1.2 Measured thermal conductivity of the sample vs filler content for graphene/PCM composites. Reproduced from Reference 23 with permission from American Chemical Society.

1.2.3 Metal Foam PCMs

Phase changing materials combined with metal foams have attracted recent attention due to the improvement in the thermal properties of the base materials. Thermal transport, effective thermal conductivity, forced and natural convection, thermal radiation as well as pool and flow boiling heat transfer in the material can be increased by incorporating it with metal foams [24]. Use of aluminum foam embedded paraffin wax was studied for heat transfer enhancement [25]. The experiments showed that the metal foam was able to improve solid liquid phase change heat transfer, due to the enhanced thermal conduction inside the metal foam. It was also evident that the structure of the metal foams affected the overall heat conduction behavior of the material.

Use of copper metal foam for passive thermal management using paraffin wax was reported in another study [26]. The apparatus consisted of a hybrid heat sink with parallel fins attached to it and copper foam-paraffin composite material saturated in its hollow basement. For comparison, improvement of thermal properties using pure paraffin wax and solid copper basement were also analyzed. The use of metal foam in the paraffin matrix reduced the time required for the wax to reach its melting point and also effectively reduced the surface temperature of the system. Surface temperature could be further reduced by decreasing the pore density and porosity of the foam. Enhancement in thermal conduction by the metal foam surpassed the natural convection in the pure paraffin wax which caused a linear increase in the temperature during the melting process.

Embedding ultrathin graphite foams into PCMs can enhance the thermal conductivity by almost 18 times without much effect on the melting temperature or mass specific heat of fusion [27]. Wide applicability of such foams to a range of PCMs can help in achieving hybrids with high thermal cycling stability.

In another study, melting phase change heat transfer on open celled metallic foams with paraffin was analyzed both theoretically and experimentally [28]. Highly porous copper metal foam was used as the nanofiller in the paraffin matrix. The inner temperature distribution as well as wall temperature distribution of the copper foam were analyzed during the melting of the paraffin wax. Local natural convection inside the wax was suppressed by the presence of the metal foam even though the thermal conductivity was increased. Numerical study was effective in predicting the velocity, interface location during phase change and thermal field of the synthesized composites. Another study also utilized PCMs stabilized with copper metal foam for safety applications inside Li ion batteries [29]. The latent heat of absorption and natural convection inside the PCMs could be used for the surface temperature reduction and maintenance of constant temperature range in the battery material. The increase in porosity and porous density, however, caused an enhancement in the surface temperature of the battery.

1.3 Classification of PCMs Based on Morphology

PCMs show different morphologies based on the synthesis method, base material and filler used during composite formation. As mentioned earlier, numerous synthesis methods have been developed to generate PCMs with different morphologies, such as electrospinning, microencapsulation, vacuum impregnation, simple submersion, etc. [30-32]. The morphologies include film, core-shell, fibrous, porous, capsules, etc. Morphology of the PCMs is an important factor, especially when these are integrated into different systems. PCMs with fibrous and porous morphology are analyzed further in this section.

1.3.1 Fibrous Structure

PCMs with fibrous morphology are synthesized mainly using electrospinning, microencapsulation and sol-gel methods. McCann *et al.* [33] used melt coaxial electrospinning, which is an innovative method for the synthesis of PCM nanofibers. This method has been successful in electrospinning organic materials like paraffin wax and creating fibers with long chain hydrocarbon in the core, in a single step. The authors focused on materials that underwent phase change near room temperature as well as body temperature, thus, making them more attractive for practical applications. Rapid solidification of hydrocarbons by the evaporative cooling of carrier solution resulted in segmented morphology inside the core of the composite material. Figure 1.3 shows scanning (SEM) and transmission images of electrospun octadecane@TiO_2-PVP nanofibers using melt coaxial spinneret.

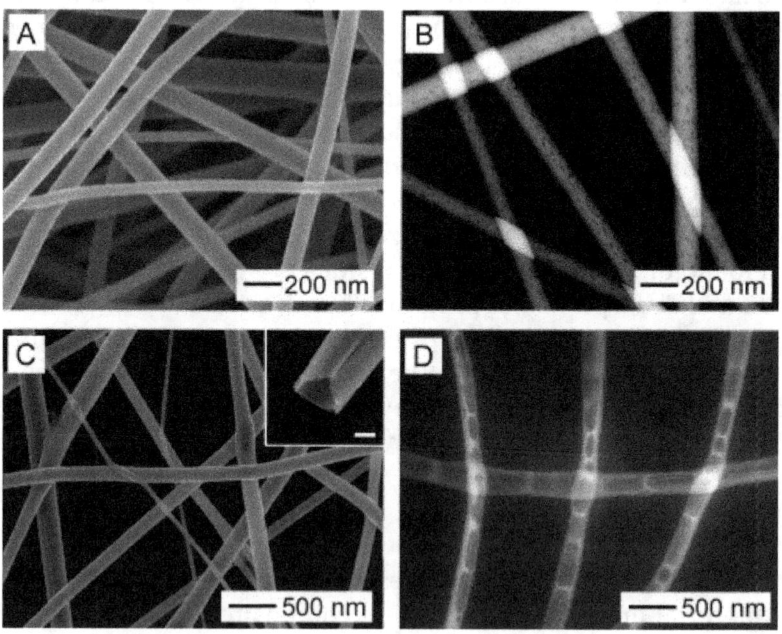

Figure 1.3 SEM image of the nanofibers with 7% octadecane loading, (B) TEM image of the fibers with the octadecane having been removed by soaking in hexane for 24 h, (C) SEM image of octadecane@TiO_2-PVP nanofibers with 45% octadecane loading and (D) TEM image of these fibers after the octadecane had been removed by soaking in hexane for 24 h. Reproduced from Reference 33 with permission from American Chemical Society.

Chen *et al.* [34] also reported electrospun fatty acids/polyethylene terephthalate (PET) composite fibers as novel form-stable phase change materials. The latent heat and phase transition temperature of the fatty acid/PET composite fibers were observed to vary with the type of the fatty acids employed in the study. In addition, these characteristics were maintained after multiple heating-cooling cycles.

Phase changing thermal materials with fibrous structure have been used in a wide range of applications, as also obvious from a number of patents on this subject [35-39]. For example, phase changing paraffinic hydrocarbons have been incorporated with different types of coatings for application in thermal insulations [38]. In another study, a phase change material from a mixture of plant oils was used for incorporation into electrospun poly(vinyl alcohol) nanofibers [40]. The authors used an emulsion electrospinning technique for this purpose.

1.3.2 Porous Structure

Vacuum impregnation method has been widely used for the synthesis of PCMs with porous structure or porous filler material. This method is (a) convenient to use, (b) economical and (c) able to load the porous fillers into the matrix effectively. The granular filler material can be anything from fibrous graphene, clay materials, fly ash, etc., and the phase changing materials can be paraffin fax, fatty acids, thickened latex, etc. [41].

Polyethylene glycol was impregnated into the porous structure of diatomite for thermal energy storage applications [42]. PEG exhibited good retention capability into the porous diatomite without any apparent leakage from the composite material. The composite showed good thermal reliability, cycling stability and melting temperature around ambient conditions. Thermal conductivity of the composite could be further improved by adding expanded graphite to the porous structure in varying mass fractions. In another study, erythritol was impregnated in different porous materials like perlite, gamma alumina and diatom earth for heat recovery applications [43]. Porosity of the materials affected the melting temperature of the phase change composite. Materials with low porosity showed lower melting point. Expanded perlite was found to be the best filler material which exhibited 83% retention of latent heat of erythritol, its complete impregnation into the pores and good cycling retention ability. Numerical investigation of the melting process of a phase change material inside a porous matrix also confirmed that the use of materials with high pore density and thermal conductivity improved the properties of PCMs for energy storage applications [44,45]. Isothermal phase changes in metallic systems using the melting of pure gallium in a porous cavity has also been numerically analyzed [46].

Zhang *et al.* [47] studied the phase changing behavior of capric acid and paraffin with three different porous building materials. It was observed that different phase change materials acted differently in the porous matrices. A number of other studies have reported phase change materials with porous expanded graphite and porous carbon materials [48,49].

1.4 Applications of PCMs

PCMs have been used in a wide range of applications like energy storage [50,51], construction [52-55], textile industry, heat storage [56-60], random access memory technology [61], data storage [62], non-volatile memories [63], etc. The properties of the PCMs to store heat energy in its latent form and to undergo changes in the phase by virtue of heat release and heat storage are made use of in most of their applications. When the temperature is supplied, it causes breaking of chemical bonds, thereby, causing a change from solid to liquid form. There are specific chemical, kinetic, thermal and physical properties that are desired for specific applications. The physical properties correspond to the attainment of desired phase equilibrium which prevents phase segregation. The thermal property is the phase change in a suitable temperature range, depending on external conditions. From the kinetic point of view, there should not be any super-cooling. From the chemical perspective, the materials should have chemical stability, non-toxicity, fire safety, compatibility and environment suitability, even after multiple cycles of use [64].

1.4.1 Energy Storage

Renewable energy is one of the major topics of discussion in 20th century. Direct solar radiation is one of the upcoming solutions for the energy crises we are facing these days with greenhouse gas emissions and other environmentally polluting energy usage. Appropriate conversion of energy has been a major research focus in the recent years. New advanced energy storage systems are also needed to be reliable, environment friendly and efficient in performance. The PCMs with improved properties are one of the attractive candidates for new generation green energy store houses. Their large scale production and application in real life situations is one of the major concerns that needs to be tackled [65].

Thermal energy storage has been mainly achieved using fatty acids and their composite PCMs. It can be utilized in space heating and other practical applications [66]. Polymer based PCMs have been used extensively used in energy storage applications. In one such study, the system consisting of polyethylene/wood flour mixed with micro-encapsulated paraffin (MEP) was developed [67]. Thermal conductivity of the system was improved by adding micro mist graphite (MMG). It was observed from the SEM images that the MEP particles remain undamaged, and form-stable PCMs had homogenous composition. Also, the leakage of molten paraffin was prevented by using the matrix and shell of MEP. It was noted that as-prepared form-stable PCMs had freezing and melting temperatures as well as latent heat suitable for latent heat thermal energy storage (LHTES) applications. It was also observed that after subjecting to 100 melt-freeze cycles, the resultant material had a better thermal stability. The form-stable PCM showed an enhancement of 17% in the thermal conductivity with the addition of 8.8 wt% MMG. The incorporation of MMG had no negative impact on the mechanical behavior of form-stable composite PCMs.

As mentioned earlier, Li *et al.* [29,68] also reported a passive thermal management system for lithium ion batteries using porous metal foam saturated with phase change materials. The composite reduced the surface temperature of the battery and correspondingly enhanced the temperature distribution owing to the increased effective thermal conductivity.

1.4.2 Smart Buildings

Construction and building materials with smart properties compared to the conservative building materials are in increasing demand now a days. Many smart building materials with energy storage capabilities have been developed using suitable phase change materials. This is achieved by incorporating PCMs into materials used for floors, walls and ceilings [69-75].

Microencapsulation is one of the methods used for the synthesis of materials for application in construction industry. For this purpose, PCMs have been incorporated with gypsum, concrete, slurries, etc. [76]. A mixture of PCMs and fluids like water or slurries can act as a two phase latent heat storage fluid. The slurries combined with PCMs are of immense use in civil applications for maintaining temperature inside the buildings [77].

Good performance stability, suitable phase transformation temperature range and high latent heat were exhibited by composites generated by blending dodecanol, butyl stearate and RT20 with organically modified montmorillonite (MMT) [78]. The composite materials showed high compatibility with gypsum. The composites were effective in decreasing the energy consumption by the buildings by decreasing the variation in temperature inside the building. RT20/MMT composite was found to the most suitable material, and the performance increased with enhancing the mass ratio of the composite. In another study, wall boards made from PCM copolymer were observed to enhance the temperature comfort inside the buildings [79]. The wall boards were successful in promoting natural convection inside the room. A numerical method was also suggested by the authors for future building applications.

1.4.3 Textile Applications

Another area where PCMs have found wide range of applications is the production of smart textiles. In these applications, these materials provide appropriate environment for the body and ensure that the physical conditions around the human body are suitable for survival or comfort [80-82].

The phase changing materials can create a small micro-climate around the body which can protect it from the changes in outside temperature. Thus, these materials help in maintaining the thermal balance inside the body. The functions of a PCM material inside a fabric range from thermal conduction as well as heat transfer between the body and the environment to the thermal isolation of the body. Thermal insulation can be achieved by using active or passive insulating materials [83]. In a representative study, Xiao *et al.* [84] developed a silk fabric with thermal controlling properties, using temperature-controlled water vapor annealing.

In-situ polymerization is a common method for the microencapsulation of PCMs for textile applications. Sarier and Onder [85] reported polyurea-formaldehyde microcapsules containing different waxes, using *in-situ* polymerization method. In another study, Sanchez *et al.* [86] reported the preparation of microcapsules with a high amount of PRS paraffin wax, prepared using Shirasu porous glass (SPG) emulsification method and simultaneous suspension polymerization, for fabric applications. The microcapsule material exhibited a high latent heat of 102.42 J/g, confirming its suitability for textile applications. Figure 1.4 shows the schematic of the microencapsulation system.

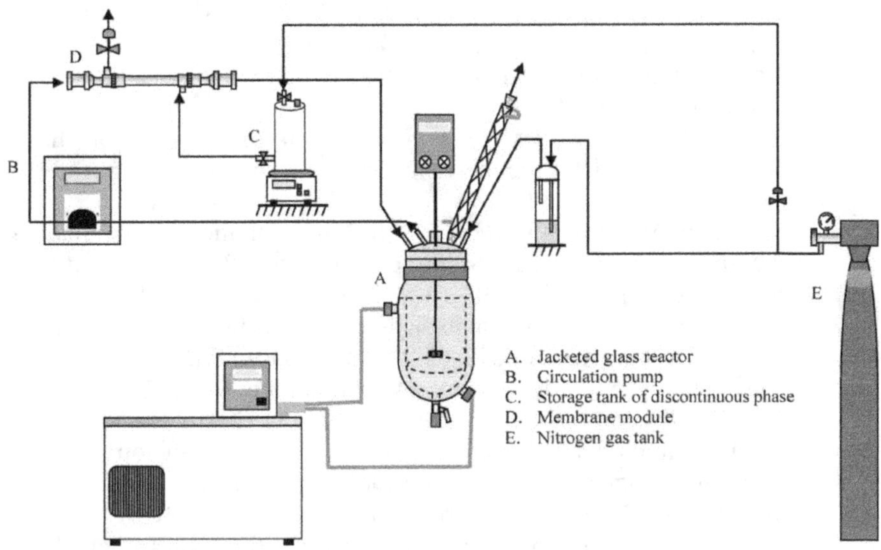

Figure 1.4 Schematic diagram of the microencapsulation system. Reproduced from Reference 86 with permission from American Chemical Society.

1.5 Conclusions

This chapter reviews the properties and applications of different phase changing materials. A number of methods have been employed to synthesize composite assemblies incorporating PCMs, however, economical, environmental friendly and bulk production methods need to be explored further. In addition, the morphology of the phase change materials needs to be taken into consideration while using these synthesis methods as it affects the material's integration for different applications. Thermal and chemical stability as well as conductivity of PCMs also need to be optimized for their practical use. Overall, PCMs are effective in taking human comfort to the next level by contributing to the development of smart buildings, energy storage media, textiles, etc.

References

1. Wang, X., Zhang, Y.-P., Xiao, W., Zeng, R.-L., Zhang, Q.-L., and Di, H.-F. (2009) Review on thermal performance of phase change energy storage building envelope. *Chinese Science Bulletin*, **54**, 920-928.
2. Shi, J.-N., Ger, M.-D., Liu, Y-.-M., Fan, Y.-C., Wen, N.-T., Lin, C.-K., and Pu, N.-W. (2013) Improving the thermal conductivity and shape-stabilization of phase change materials using nanographite additives. *Carbon*, **51**, 365-372.
3. Jankowski, N. R., and McCluskey, F. P. (2014) A review of phase change materials for vehicle component thermal buffering. *Applied Energy*, **113**, 1525-1561.
4. Ma, Z., Lin, W., and Sohel, M. I. (2016) Nano-enhanced phase change materials for improved building performance. *Renewable and Sustainable Energy Reviews*, **58**, 1256-1268.
5. Parameshwaran, R., and Kalaiselvam, S. (2014) Energy conservative air conditioning system using silver nano-based PCM thermal storage for modern buildings. *Energy and Buildings*, **69**, 202-212.
6. Khodadadi, J. M., Fan, L., and Babaei, H. (2013) Thermal conductivity enhancement of nanostructure-based colloidal suspensions utilized as phase change materials for thermal energy storage: A review. *Renewable and Sustainable Energy* Reviews, **24**, 418-444.
7. Kashani, S., Ranjbar, A. A., Madani, M. M., Mastiani, M., and Jalaly, H. (2013) Numerical study of solidification of a nano-enhanced phase change material (NEPCM) in a thermal storage system. *Journal of Applied Mechanics and Technical Physics*, **54**, 702-712.
8. Wu, S. Y., Wang, H., Xiao, S., and Zhu, D. S. (2013) An investigation of melting/freezing characteristics of nanoparticle-enhanced phase change materials. *Journal of Thermal Analysis and Calorimetry*, **110**, 1127-1131.
9. Appavoo, K., Wang, B., Brady, N. F., Seo, M., Nag J., Prasankumar R. P., Hilton D. J., Pantelides, S. T., and Haglund, Jr. R. F., (2014) Ultrafast phase transition via catastrophic phonon collapse driven by plasmonic hot-electron injection. *Nano Letters*, **14**(3), 1127-1133.
10. Chen, L., Zou, R., Xia, W., Liu, Z., Shang, Y., Zhu, J., Wang, Y., Lin, J., Xia, D., and Cao, A. (2012) Electro- and photodriven phase changec based on wax-infiltrated carbon nanotube sponges. *ACS Nano*, **6**(12), 10884-10892.
11. Tumirah, K., Hussein, M. Z., Zulkarnain, Z., and Rafeadah, R. (2014) Nano-encapsulated organic phase change material based on copolymer nanocomposites for thermal energy storage. *Energy*, **66**, 881-890.
12. Fang, G., Chen, Z., and Li, H. (2010) Synthesis and properties of microencapsulated paraffin composites with SiO_2 shell as thermal en-

ergy storage materials. *Chemical Engineering Journal*, **163**, 154-159.

13. Fang, G., Li, H., Yang, F., Liu, X., and Wu, S. (2009) Preparation and characterization of nano-encapsulated n-tertradecane as phase change material for thermal energy storage. *Chemical Engineering Journal*, **153**, 217-221.
14. Wu, W., Bostanci, H., Chow, L. C., Ding, S. J., Hong, Y., Su, M., Kizito, J. P., Gschwender, L., and Snyder, C. E. (2011) Jet impregnation and spray cooling using slurry of nano-encapsulated phase change materials. *International Journal of Heat and Mass Transfer*, **54**, 2715-2723.
15. Karan, N. S., Suresh, S., Sarma, D. D., Paromita, K., Ravishankar, N., and Pradhan, N. (2011) Thermally controlled cyclic insertion/ejection of dopant ions and reversible zinc blende/wurtzite phase changes in ZnS nanostructures. *Journal of The American Chemical Society*, **133**, 1666-1669.
16. Ettouney, H., El-Dessouky, H., and Al-Kandari, E. (2004) Heat transfer characteristics during melting and solidification of phase change energy storage process. *Industrial and Engineering Chemistry Research*, **43,** 5350-5357.
17. Feng, L., Zhao, W., Zheng, J., Frisco, S., Song, P., and Li, X. (2011) The shape-stabilized phase change materials composed of polyethylene glycol and various mesoporous matrices (AC, SBA-15 and MCM-41). *Solar Energy Materials and Solar Cells*, **95**(12), 3550-3556.
18. Shaikh, S., Lafdi, K., and Hallinan, K. (2008) Carbon nano-additives to enhance latent energy storage of phase change materials. *Journal of Applied Physics*, **103**, 094302.
19. Wang, J., Xie, H., and Xin, Z. (2009) Thermal properties of paraffin based composites containing multi-walled carbon nanotubes. *Thermochimica Acta*, **488**(1-2), 39-42.
20. Elgafy, A., and Lafdi K. (2005) Effect of carbon nanofiber additives on thermal behavior of phase change materials. *Carbon*, **43**, 15, 3067-3074.
21. Bonnet, P., Sireude, D., Garnier, B., and Chauvet, O. (2007) Thermal properties and percolation in carbon nanotube-polymer composites. *Applied Physics Letters*, **91**, 201910.
22. Koga, K., Gao, G. T., Tanaka, H., and Zeng, X. C. (2001) Formation of ordered ice nanotubes inside carbon nanotubes. *Nature*, **412**, 802-805.
23. Yavari, F., Fard, H. R., Pashayi, K., Rafiee, M. A., Zamiri, A., Yu, Z., Ozisik, R., Borca-Tasciuc, T., and Koratkat, N. (2011) Enhanced thermal conductivity in a nanostructured phase change composite due to low concentration graphene additives. *The Journal of Physical Chemistry C*, **115**, 8753-8758.
24. Zhao, C. Y. (2012) Review on thermal transport in high porosity cellular metal foams with open cells. *International Journal of Heat and*

Mass Transfer, **55**(13), 3618-3632.

25. Chen, Z., Gao, D., and Shi, J. (2014) Experimental and numerical study on melting of phase change materials in metal foams at pore scale. *International Journal of Heat and Mass Transfer*, **72**, 646-655.
26. Qu, Z. G., Li, W. Q., Wang, J. L., and Tao, W. Q. (2012) Passive thermal management using metal foam saturated with phase change material in a heat sink. *International Communications in Heat and Mass Transfer*, **39**(10), 1546-1549.
27. Ji, H., Sellan, D. P., Pettes, M. T., Kong, X., Ji, J., Shi, L., and Ruoff, R. S. (2014) Enhanced thermal conductivity of phase change materials with ultrathin-graphite foams for thermal energy storage. *Energy and Environmental Science*, **7**, 1185-1192.
28. Li, W. Q., Qu, Z. G., He, Y. L., and Tao, Y. B. (2012) Experimental and numerical studies on melting phase change heat transfer in open-cell metallic foams filled with paraffin. *Applied Thermal Engineering*, **37**, 1-9.
29. Li, W. Q., Qu, Z. G., He, Y. L., and Tao, Y. B. (2014) Experimental study of a passive thermal management system for high-powered lithium ion batteries using porous metal foam saturated with phase change materials. *Journal of Power Sources*, **255**(1), 9-15.
30. Fukai, J., Hamada, Y., Morozumi, Y., and Miyatake, O. (2003) Improvement of thermal characteristics of latent heat thermal energy storage units using carbon-fiber brushes: experiments and modeling. *International Communications in Heat and Mass Transfer*, **46**, 4513-4525.
31. Chow, L. C., Zhong, J. K., and Beam, J. E. (1996) Thermal conductivity enhancement for phase change storage media. *International Communications in Heat and Mass Transfer*, **23**(1), 91-100.
32. Cui, Y., Liu, C., Hu, S., and Yu, X., (2011) The experimental exploration of carbon nanofiber and carbon nanotube additives on thermal behavior of phase change materials. *Solar Energy Materials and Solar Cells*, **95**(4), 1208-1212.
33. McCann, J. T., Marquez, M., and Xia, Y. (2006) Melt coaxial electrospinning: A versatile method for the encapsulation of solid materials and fabrication of phase change nanofibers. *Nano Letters*, **6**(12), 2868-2872.
34. Chen, C., Wang, L., and Huang, Y. (2008) Morphology and thermal properties of electrospun fatty acids/polyethylene terephthalate composite fibers as novel form-stable phase change materials. *Solar Energy Materials and Solar Cells*, **92**(11), 1382-1387.
35. Shannon, T. (2003) Fibrous Materials Exhibiting Thermal Change During Use, patent US 20050136765A1.
36. Pause, B. (1997) Interactive Thermal Insulating System having a Layer Treated with a Coating of Energy Absorbing Phase Change Material Adjacent a Layer of Fibers Containing Energy Absorbing

Phase Change Material, patent US 6077597A.
37. Buckley, T. M. (1992) Flexible Composite Material with Phase Change Thermal Storage, patent US 6004662A.
38. Bryant, Y. G., Colvin, D. P., Driscoll, J. C., and Mulligan, J. C. (1995) Thermal Insulating Coating Employing Microencapsulated Phase Change Material and Method, patent US 5804297A.
39. Bryant, Y. G., and Colvin, D. P. (1987) Fiber with Reversible Enhanced Thermal Storage Properties and Fabrics Made Therefrom, patent US 4756958A.
40. Zdraveva, E., Fang, J., Mijovic, B., and Lin, T. (2015) Electrospun Poly (vinyl alcohol)/ phase change material fibers: morphology, heat properties, and stability. *Industrial and Engineering Chemistry Research,* **54**, 8706-8712.
41. Zhang, D., Zhou, J., Wu, K., and Li, Z. (2005) Granular phase changing composites for thermal energy storage. *Solar Energy,* **78**(3), 471-480.
42. Karaman, S., Karaipekli, A., Sarı, A., and Biçer, A. (2011) Polyethylene glycol (PEG)/diatomite composite as a novel form-stable phase change material for thermal energy storage. *Solar Energy Materials and Solar Cells,* **95**(7), 1647-1653.
43. Nomura, T., Okinaka, N., and Akiyama, T. (2009) Impregnation of porous material with phase change material for thermal energy storage. *Materials Chemistry and Physics,* **115**(2), 846-850.
44. Mesalhy, O., Lafdi, K., Elgafy, A., and Bowman, K. (2005) Numerical study for enhancing the thermal conductivity of phase change material (PCM) storage using high thermal conductivity porous matrix. *Energy Conversion and Management,* **46**(6), 847-867.
45. Lafdi, K., Mesalhy, O., and Elgafy, A. (2008) Graphite foams infiltrated with phase change materials as alternative materials for space and terrestrial thermal energy storage applications. *Carbon,* **46**(1), 159-168.
46. Brent, A. D., Voller, V. R., and Reid, K. J. (1988) Enthalpy-porosity technique for modeling convection-diffusion phase change: application to the melting of a pure metal. *Numerical Heat Transfer,* **13**(3), 297-318.
47. Zhang, D., Tian, S., and Xiao, D. (2007) Experimental study on the phase change behavior of phase change material confined in pores. *Solar Energy,* **81**(5), 653-660.
48. Zhang, Z., and Fang, X. (2006) Study on paraffin/expanded graphite composite phase change thermal energy storage material. *Energy Conversion and Management,* **47**(3), 303-310.
49. Zhou, D., and Zhao, C. Y. (2011) Experimental investigations on heat transfer in phase change materials (PCMs) embedded in porous materials. *Applied Thermal Engineering,* **31**(5), 970-977.
50. Farid, M., and Kong, W. J. (2001) Underfloor heating with latent heat

storage. *Proceedings of the Institution of Mechanical Engineers, Part A, Journal of Power and Energy*, **215**, 601-609.
51. Farkas, D., and Birchenall, C. E. (1985) New eutectic alloys and their heats of transformation. *Metallurgical and Materials Transactions A*, **16**, 323-328.
52. Vafai, K., and Sarkar, S. (1986) Condensation effects in a fibrous insulation slab. *Journal of Heat Transfer*, **108** (8), 667-675.
53. Kedl, R. J. (1990) Conventional Wallboard with Latent Heat Storage for Passive Solar Applications. *Proceedings of the 25th Intersociety Energy Conversion Engineering Conference*, USA, doi: 10.1109/IECEC.1990.716492.
54. Feldman, D., Banu, D., Hawes, D., and Ghanbari, E. (1991) Obtaining an energy storing building material by direct incorporation of an organic phase change material in gypsum wallboard. *Solar Energy Materials*, **22**, 231-242.
55. Drake, J. B. (1987) *A study of the Optimal Transition Temperature of PCM Wallboard for Solar Energy Storage*. Oak Ridge National Laboratory, USA. Online: https://www.osti.gov/servlets/purl/6100261 [accessed 21st December 2018].
56. Charach, C., Zarmi, Y., and Zemel, A. (1987) Simple method for assessing the thermal performance of PCM panels. *Proceedings of the Biennial Congress of the International Solar Energy Society*, Germany, pp. 1212-1216.
57. Pincemin, S., Olives, R., Py, X, and Christ M. (2008) Highly conductive composites made of phase change materials and graphite for thermal storage. *Solar Energy Materials & Solar Cells*, **92**, 603-613.
58. Ait Adine, H., and El Qarnia, H. (2009) Numerical analysis of the thermal behaviour of a shell-and-tube heat storage unit using phase change materials. *Applied Mathematical Modelling*, **33**, 2132-2144.
59. Lacroix, M. (1993), Numerical simulation of a shell and tube latent heat thermal energy storage unit. *Solar Energy*, **50**(4), 357-367.
60. Dincer, I., and Rosen, M. A. (2002) *Thermal Energy Storage, Systems and Applications*, John Wiley & Sons, USA.
61. Raoux, S., Shelby, R. M., Jordan-Sweet, J., Munoz, B., Salinga, M., Chen, Y.-C., Shih, Y.-H., Lai, E.-K., and Lee, M.-H. (2008) Phase change materials and their application to random access memory technology. *Microelectronic Engineering*, **85**(12), 2330-2333.
62. Lencer, D., Salinga, M., and Wuttig, M. (2011) Design rules for phase-change materials in data storage applications. *Advanced Materials*, **23**, 2030-2058.
63. Raoux, S., Welnic, W., and Ielmini, D. (2010), Phase change materials and their application to nonvolatile memories. *Chemical Reviews*, **110** (1), 240-267.
64. Baetens, R., Jelle, B. P., and Gustavsen, A. (2010) Phase change materials for building applications: A state-of-the-art review. *Energy*

and Buildings, **42**(9), 1361-1368.

65. Akhiani, A. R., Mehrali, M., Latibari, S. T., Mehrali, M., Mahlia, T. M. I., Sadeghinezhad, E., and Metsellar, H. S. C. (2015) One-step Preparation of form-stable phase change material through self-assembly of fatty acid and graphene. *The Journal of Physical Chemistry C*, **119**(40), 22787-22796.
66. Sharma, A, Tyagi, V. V., Chen, C. R, and Buddhi, D. (2009) Review on thermal energy storage with phase change materials and applications. *Renewable and Sustainable Energy Reviews*, **13**(2), 318-345.
67. Li, J., Xue, P., Ding, W., Han, J., and Sun, G. (2009) Micro-encapsulated paraffin/high-density polyethylene/wood flour composite as form-stable phase change material for thermal energy storage. *Solar Energy Materials and Solar Cells*, **93**(10), 1761-1767.
68. Li, W. Q., Wan, H., Jing, T. T., Li, Y. B., Liu, P. J., He, G. Q., and Qin, F. (2019) Microencapsulated phase change material (MEPCM) saturated in metal foam as an efficient hybrid PCM for passive thermal management: A numerical and experimental study. *Applied Thermal Engineering*, **146**, 413-421.
69. Khudhair, A. M, and Farid, M. M. (2004) A review on energy conservation in building applications with thermal storage by latent heat using phase change materials. *Energy Conversion and Management*, **45**(2), 263-275.
70. De Gracia, A., and Cabeza, L. F. (2015) Phase change materials and thermal energy storage for buildings. *Energy and Buildings*, **103**, 414-419.
71. Heim, D., and Clarke, J. A. (2003) Numerical Modeling and Thermal Simulation of Phase Change Material with ESP-R. E*ighth international IBPSA Conference*, The Netherlands, pp. 459-466.
72. Tyagi, V. V., Kaushik, S. C., Tyagi, S. K., and Akiyama, T. (2011) Development of phase change materials based microencapsulated technology for buildings: a review. *Renewable and Sustainable Energy Reviews*,**15**, 1373-1391.
73. Ibanez, M., Lazaro, A., Zalba, B., and Cabeza, L. F. (2005) An approach to the simulation of PCMs in building applications using TRNSYS. *Applied Thermal Engineering*, **25**, 1796-807.
74. Boh, B., and Sumiga, B. (2008) Microencapsulation technology and its applications in building construction materials. *RMZ - Materials and Geo environment*, **55**(3), 329-344.
75. Chandra Sekhar, G., Venkatesu, P., Hofman, T., and Prabhakara Rao, M. V. (2002) Solid-liquid equilibria of long chain n-alkanes (C18-C28) in n,n-dimethylacetamide. *Fluid Phase Equilibrium*, **201** 219-231.
76. Morikawa, J., and Hashimoto, T. (2000) Simultaneous measurement of heat capacity and thermal diffusivity in solid-solid and solid-liquid phase transitions of n- alkane. *Thermochimica Acta*, **352-353**,

291-296.

77. Delgado, M., Lazaro, A., Mazo, J., and Zalba, B. (2012) Review on phase change material emulsions and microencapsulated phase change material slurries: Materials, heat transfer studies and applications. *Renewable and Sustainable Energy Reviews*, **16**(1), 253-273.
78. Fang, X., Zhang, Z., and Chen, Z. (2008) Study on preparation of montmorillonite-based composite phase change materials and their applications in thermal storage building materials. *Energy Conversion and Management*, **49**(4), 718-723.
79. Kuznik, F., and Virgone, J. (2009) Experimental assessment of a phase change material for wall building use. *Applied Energy*, **86**(10), 2038-2046.
80. Alvarado, J. L., Marsh, C., Sohn, C., Vilceus, M., Hock, V., Phetteplace, G, and Newell, T. (2006) Characterization of supercooling suppression of microencapsulated phase change material by using DSC. *Journal of Thermal Analysis and Calorimetry*, **86** (2) 505-509.
81. Quanying, Y., Chen, L., and Lin, Z. (2008) Experimental study on the thermal storage performance and preparation of paraffin mixtures used in the phase change wall. *Solar Energy Materials and Solar Cells*, **92,** 1526-1532.
82. Van Miltenburg, J. C. (2000) Fitting the heat capacity of liquid n-alkanes: new measurements of n-heptadecane and n-octadecane. *Thermochimica Acta*, **343,** 57-62.
83. Mondal, S. (2008) Phase change materials for smart textiles - An overview. *Applied Thermal Engineering*, **28,** 1536-1550.
84. Hu, X., Shmelev, K., Sun, L., Gil, E.-S., Park, S.-H., Cebe, P., and Kaplan, D. L. (2011) Regulation of silk material structure by temperature-controlled water vapor annealing. *Biomacromolecules*, **12**, 1686-1696.
85. Sarier, N., and Onder, E. The manufacture of microencapsulated phase change materials suitable for the design of thermally enhanced fabrics. *Thermochimica Acta*, **452**(2), 149-160.
86. Sanchez, L., Lacasa, E., Carmona, M., Rodrigues, J. F., and Sanchez, P. (2008) Applying an experimental design to improve the characteristics of microcapsules containing phase change materials for fabric uses. *Industrial and Engineering Chemistry Research*, **47**(23), 9783-9790.

2

Phase Change Materials for Solar Energy Applications

Hoda Akbari,[a] David G Peña,[a] Rebeca S Pizarro,[a] Hind Ahmed,[a] Maria C. Browne,[a] Chuka O'koli,[a] Edward Guionneau,[a] Ming Jun Huang[b] and Sarah J. McCormack[a,c,*]

[a]*Department of Civil, Structural and Environmental Engineering, Trinity College, Dublin 2, Ireland*
[b]*School of Architecture and the Built Environment, Built Environment Research Institute, Ulster University, Belfast, UK*
[c]*Dublin Energy Lab., Dublin Institute of Technology, Grangegorman, Dublin 7, Ireland*

**Corresponding author*: mccorms1@tcd.ie

2.1 Introduction

Increasing levels of greenhouse gas (GHG) emissions and energy dependency on other countries have brought governments to develop and improve various ways of producing energy. There are many different methods of renewable production of energy. Currently, solar photovoltaics are one of the most popular methods and rapidly increasing with a worldwide growth of 60% over 5 years from 2007 to 2012 [1]. This is due to fact that solar panels are simply designed products that require low maintenance and are easy to install.

Photovoltaic systems can be installed on nearly all terrain permitting easy distribution of the energy produced among a community or country assisting resolving local grid issues. Furthermore, they also allow energy independence on a local or even national scale, which is a major geopolitical issue for many countries.

Although, before 2010 the electricity produced by a PV system was much more expensive than that produced by wind and geothermal power and up to 6 times more expensive than electricity produced by coal and gas [2], presently the PV cost has decreased 4-5 times becoming a real alternative in the close future to the fossil fuels [3].

Phase Change Materials, edited by Vikas Mittal

This high cost of PV energy is due to the high cost of PV panels and their low level of efficiency - they only convert 15% to 20% of solar radiation into electricity energy.

The applications for photovoltaic (PV) systems differ and they include field, transport and building applications. Over the decade, the incorporation of photovoltaics as an integrated part of buildings has exponentially increased. They are referred to as building integrated photovoltaic systems (BIPV) and have one of the fastest growing markets globally [4]. BIPV systems possess many advantages and are summarized as follows [5]:

I. When PV systems are integrated, unified or mounted onto buildings, they mitigate against the cost of acquiring land and associated costs like fencing and cost of support structures which are required for ground installation of the PV modules.
II. The overall cabling costs can be significantly reduced because the buildings are typically connected to the grid.
III. The losses and cost that are correlated with the transmission and distribution of electricity are mitigated against, because the electricity is being originated close to the point of use.
IV. The integration of PV into buildings displace the use of other building materials while performing structural duties like weather protection.

Due to PV cells being the most expensive component of a photovoltaic system, the idea of concentrating solar radiation into a smaller area of cells gave prospect into reduction of the system overall cost [6]. The reduction in the BIPV system can be achieved either through the increase of the system efficiency or through replacement of the PV components (PV cells) with reflective material [7]. The use of the reflective or refractive material in the latter concentrates solar energy onto the PV cells allowing for an increased luminous flux on the PV cell surface [7], a reduced cell area per output and a reduced overall system cost per unit of energy produced [6].

Furthermore, the efficiency of solar cells decreases by 0.5% for every degree increase above the solar cells standard operating temperature [8]. If the trends of increasing PV electricity production are going to continue, the efficiency of PV cells need to be maintained during high temperature conditions. Phase change materials (PCMs) can absorb or release heat while they are changing phase, for instance from solid to liquid. PCMs have melting ranges at which the temperature stays constant until the material is fully liquefied or solidified; referred to as the latent heat stage. During this stage, PCMs can absorb or release large amounts of heat energy. The phase change process results in a cooling effect which has proven to be very useful at regulating temperatures in various applications. Previous experiments undertaken combining PCMs and solar panels have shown very encouraging results [9]. However, one of the main issues that was highlighted is the material would not always fully re-solidify overnight, meaning that the PCM would not have maximum capacity for use the next day.

Most solar radiation reaching a PV cell is converted into thermal energy increasing the cell temperature. Temperature increases in photovoltaic (PV) module lead to immediate efficiency losses and can accelerate long-term degradation. Silicon PV modules operated at elevated temperatures exhibit lower efficiency in converting solar energy into electrical energy. Temperature elevation is inevitable when the rate of solar heat gained by the PV is higher than the rate of heat lost to the ambient environment [10].

2.2 Phase Change Materials

PCMs are materials that have the capacity to absorb heat while they change phase. As the phase change happens in their latent heat stage, the temperature of the material stays constant as heat gets absorbed or rejected, at that point the material is in a mix phase, a mixture of solid and liquid. Heat will be absorbed or released until the material reaches its maximum heat capacity or is fully discharged at which point the material will be fully melted. Once the material is back into a single phase, the temperature will start increasing or decreasing again. Water is a good example of a PCM. The temperature of an ice cube will quickly reach 0 °C and will stay constant at that temperature until it is completely liquefied, which is the latent heat stage. As it liquefies or solidifies, heat from the atmosphere will be absorbed by the material.

In 1978, it was suggested that PCM could act as a potential thermal storage if integrated with a PV. This concept was patented in 1983 [11], but was not developed commercially. However, following this, PCM was not investigated as a means of cooling PV until the mid 1990s.

The advantages of PCMs are (i) no volume change during phase transition; (ii) increase in specific heat transfer area (large surface to volume ratio); (iii) no leakage of PCM from matrix; (iv) protection from external environment; (v) easy application; (vi) improvement in material compatibility; (vii) improved handling during production; (viii) improved cycling stability since phase separation is restricted to microscopic distances, etc. Hence, extent of enhancement increases with heat flux, which is also investigated further by comparing the materials in the systems designed for PV for CPC and other related thermal applications.

When undergoing phase transition, PCMs can provide temperature stabilization. They are used in one of two main ways: temperature control or thermal energy storage (TES) [12]. Due to the low thermal conductivity of PCMs, their categorical use for temperature control and energy storage is finite. The effect of solidification at the heat transfer surface can lead to a significant reduction in heat transfer when energy is extracted from the liquid phase [13]. A PCM heats reasonably at initial heating, but on getting to the melting/solidification temperature, there is absorption of the latent heat by the material which leads to phase change from a solid to a liquid or a liquid to a gas. As the PCM melts, its temperature stabilizes even as it continues to absorb heat during the phase change. The temperature range and time taken, over which the phase change occurs, depend on the mass and thermal conductivity of PCM and any enhanced heat transfer elements therein [14]. This process is illustrated by Gunther *et al.* [15]. PCM can yield compact thermal energy storage over a precise temperature range, and once PCM has fully changed phase, the material will begin to heat sensibly again.

Ideal PCMs must have a large latent heat of fusion, high thermal conductivity in both liquid and solid phase with no phase separation and a melting temperature lying in the practical range of operation. They should mostly melt congruently with minimum sub-cooling. They also should be chemically stable, non-toxic, non-flammable and non-corrosive. Besides that, they must also have low density variation and a high overall density. Of all these attributes, the most difficult to achieve is high thermal conductivity. Once all the characteristics are met, the PCM would ideally be cheap and abundant.

PCMs are divided into organic (paraffin, fatty acid), inorganic (hydrated salts) and eutectic (mixture of organic or inorganic PCM) within the required melting temperature of 20 °C to 100 °C. Cabeza *et al.* divided PCMs into groups [16]:

- Cooling applications up to 21 °C

- 22-28 °C for comfort in building applications
- 29-60 °C for hot water applications
- High temperature applications requiring PCM of between 61 °C and 120 °C.

Organic PCMs are characterized by presence of carbon atoms [17]; they usually have low melting temperatures (below 220°C) and are therefore generally used for room heating. Non-Paraffin compounds include esters, acids and alcohols. Organic PCMs have very low corrosion and good thermal/chemical stability but have low latent and thermal conductivity [18].

On the other hand, inorganic PCMs have greater latent heat but are worse with respect to corrosion than organic PCMs and also have weaker thermal stability. Salt hydrates, a type of inorganic PCM, are formed by water absorption by the anhydrous salt at ambient temperatures. They have a phase change enthalpy that depends on bond strength between water molecules and the salts [17,18]. For high temperature storage (above 420 °C), molten salts have been widely used which fall into the inorganic PCM category [18].

Siegel *et al.* [19] commenced a simplified analysis to determine the heat transfer enhancement provided by the inclusion of high conductivity particles in a low conductivity salt. It was assumed the particles inside the container retained a uniform distribution during both freezing and melting cycles, while they neglected sub-cooling energy. It was resolved that there was an increase in the thermal conductivity in both the melting and solidifying processes due to the embedded particles. It was observed that depending on the ratio of particle to matrix conductivity, the particles occupying 20% of the storage by volume increased the rate of heat extraction by 10% to 20%.

Cabeza *et al.* [20] initiated experimental tests to study the heat transfer enhancement of PCM for a small thermal energy storage device using stainless steel fins, copper fins or a graphite composite material. It was concluded that the addition of stainless steel fins in the PCM did not increase the heat flux considerably, and the addition of copper pieces enhanced heat transfer especially during the melting process. The highest heat transfer flux of the three methods was found using the graphite composite material which showed a heat transfer four times higher than that of the pure PCM. The paraffin used in this analysis was RUBITHERM RT 30 and it had a melting point of 30 °C. Comparisons of PCM melting and solidification with and without fins were made and it was observed that the heat transfer during solidification was greater if fins were included, and a 40% reduction in the solidification time was achieved.

Experimental tests and numerical simulations were undertaken [21] to study the thermal behavior of PV/PCM systems. The numerical simulations were validated by the experimental measurements. In the experimental work, the PCM RT25 with a melting temperature of 26.6 °C was used and comparisons of three different systems was carried out, which included a single flat aluminum plate system, a PV/PCM system without internal fins and a PV/PCM system with internal fins. All systems used in the experiment had the same sized front aluminum plate which was 0.3 m long and 0.132 m high, along with a uniform applied solar intensity of 750 W/m^2 at the front surface. Observations showed that after 60 minutes, the front surface temperature of the single plate system was constant at 63 °C. The front surface of the PV/PCM system without fins retained a temperature of around 36 °C between 50 and 100 minutes, then it slowly increased to 44 °C at 280 minutes. The front surface of the PV/PCM system with two fins maintained a temperature of around 32 to 33 °C between 50 to 100 minutes, and then sharply increased to 38 °C at 170 minutes. Conclusions were made that the fins facilitated a more uniform temperature distribution within the PV/ PCM system. Simulations further determined that for a south east orientated PV/PCM system having a 30 mm depth and no fins in the PCM container, the front surface temperature could be maintained at less than 35 °C for UK weather data during the summer solstice. The simulations were repeated for three consecutive days for confirmation [21]. The authors also conducted an experimental evaluation of PCMs for PV cooling. It was determined that RUBITHERM RT25 had an improved thermal control potential than that of RUBITHERM GR40 for the PV panel. Experimental tests on embedding different numbers and types of fins in the PCM to improve the effect of thermal management of PVs were also conducted. Huang *et al.* [22] reported the development of a three-dimensional numerical model which was used to simulate a phase change material container attached to a PV system. The PCM was used to regulate the temperature rise of the PV cells. For regulation of the temperature of PV cells, a PCM possessing suitable phase transition temperature is required. With PCM, high efficiencies of the PV cells can be maintained for an extended period of time [23,24]. The advantage of PCM is its ability to not only store heat but also allow for the heat to be used asynchronously [25]. An experimental evaluation of the effects of convection and crystalline segregation in a PCM as a function of efficiency of heat transfer within the finned PV/PCM system has been studied [26]. The thermal performances of bulk PCM with crystallization segregation for different internal fin arrangements are presented. It is noted that the addition of internal fins improves the temperature control of the PV in a PV/PCM system.

2.3 The Use of PCMs for Solar Energy Applications

The use of PCM with solar technologies has been shown to maintain the efficiency of some solar energy technologies and can also be used as an energy store which can be charged during daylight hours to be used during the night.

The availability of solar energy is limited to the central hours of the day, and there is a large variation in the intensity of solar radiation throughout the year as can be seen in Figure 2.1. Currently, there is a wide variety of technologies that harness solar energy [27]. One of the most efficient technologies is solar thermal energy systems, either through flat collectors or through complex concentration systems. In these types of installations, the objective is the heating of a fluid. In order to understand the potential applications of PCMs with

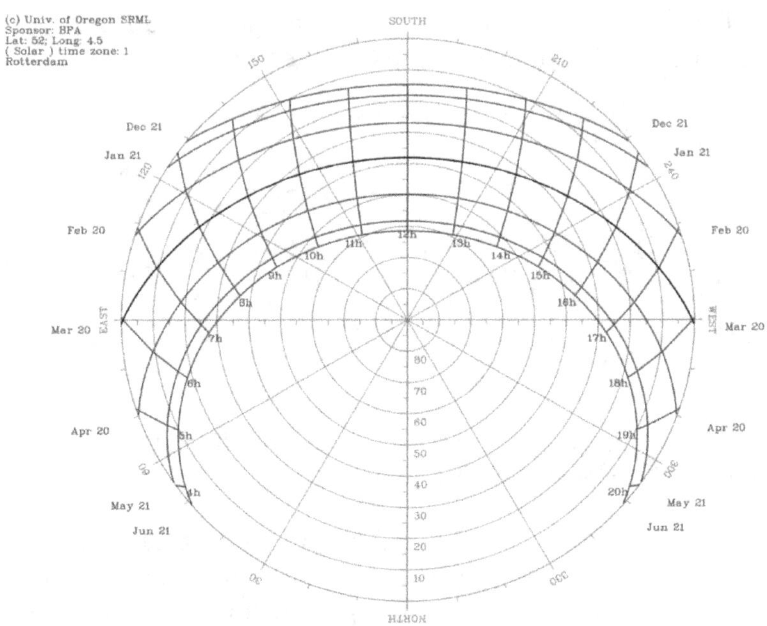

Figure 2.1 Polar diagram for Rotterdam. Sun path lines along the year.

different solar energy technologies, each of technology will be described below.

2.3.1 Solar Thermal Energy for Building and Industry

The use of low temperature solar thermal energy in homes and industries [28] is becoming increasingly widespread. These systems allow, on one hand, a saving in energy cost, as well as a reduction of CO_2 emissions as solar energy is a clean and renewable resource that can replace the use of fossil fuels. Through solar collectors, it is possible to convert solar energy into thermal energy that is transferred into a liquid or gaseous fluid that is circulated inside, usually water or air. These collectors can be divided into two categories: non-concentrating and concentrating. Domestic thermal solar energy mainly implements non-concentrating collectors that can be either flat plate or evacuated tube collectors [29]. Flat plate collectors, as shown in Figure 2.2 (left), are of simple geometry, cost efficient, robust and effective even with diffuse solar radiation. Evacuated tube collectors contain vacuum-sealed tubes with a heat pipe inside, as illustrated in Figure 2.2 (right). Convection and conduction losses are reduced by the vacuum, so it can operate at higher temperatures in comparison with flat plate collectors. Their efficiency in day-long performance is also higher at lower incidence angles due to their geometry.

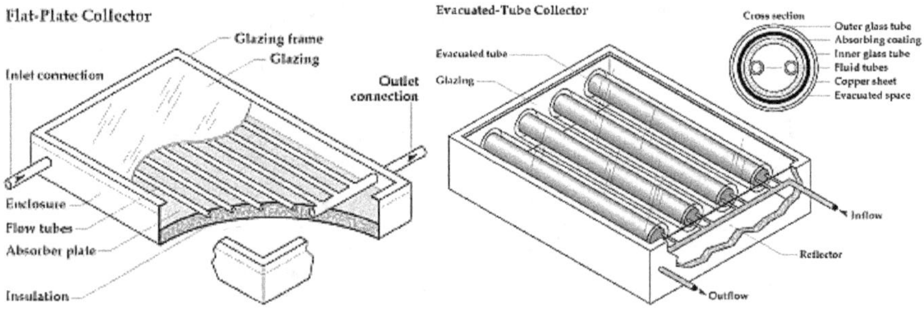

Figure 2.2 Schematic diagram of a flat plate collector (left) and evacuated tube collector (right) for solar energy (Source: U.S. Department of Energy).

However, the variability of solar radiation makes it necessary to use thermal storage systems (TES). During the central hours of the day, when solar radiation is higher, energy consumption decreases, having an excess of heat that, if it is not stored, is dissipated into the environment.

In case of having thermal energy storage (TES) systems in the installation, it is possible to use the excess energy during the peaks of energy demand or in the night. Traditionally, TES in solar thermal for domestic and industrial installations has been carried out by large tanks of fluid [30-32], usually water, where heat is stored by raising the temperature of the fluid (sensible heat). It is possible to store heat with a lower volume and greater stability of the fluid temperature through the use of PCM in these tank designs which is shown in Figure 2.3 [33-35]. These factors allow the installation costs to be lower, in addition to being able to install higher power in homes with small spaces. The wide variety of existing PCMs [31,36] allows to adjust the temperature of the phase change to the optimal operating temperature of the heating system, domestic hot water or to the industrial process.

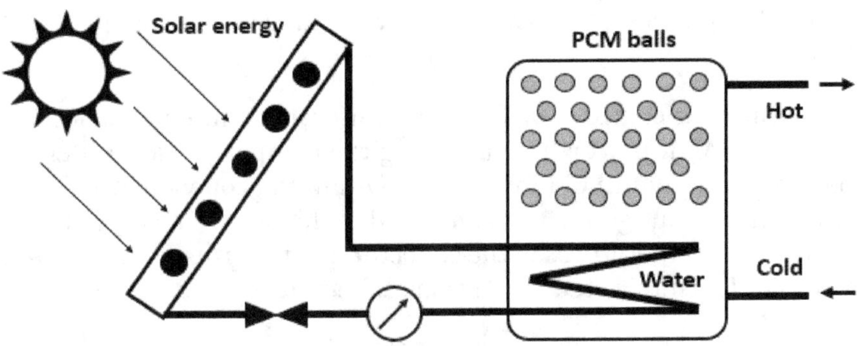

Figure 2.3 Schematic of a solar thermal installation with a TES-PCM.

The main limitation of PCMs is the low thermal conductivity of some materials. For this reason, a significant effort is being made by researchers to achieve optimized tank designs or develop new materials with higher thermal conductive properties that allow an efficient operation. The work of Kenisarin & Mahkamov [31] includes a complete review of different system design with fins or different geometries of PCM containers that improve the thermal conductivity of the system. Figure 2.4 shows an example of 4 geometries with fins, where it is verified that the most optimal geometry is represented in Figure 2.4(d) due to the higher number of fins which increase the heat transfer surface in contact with the PCM.

In the work of Sharif *et al.* [30], a detailed review of different PCM applications in domestic hot water systems is presented. In addition

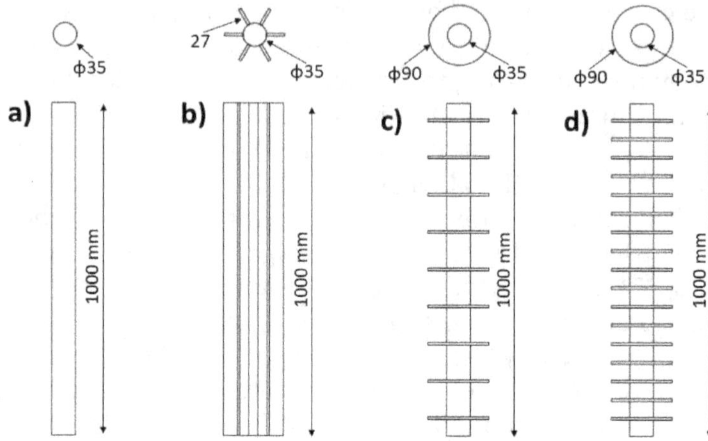

Figure 2.4 Geometries studied to increase the heat transference in a PCM system: a) single tube, b) rectangular fins, c) 17 circular fins and d) 34 circular fins.

to presenting a comparison between different tank geometries for TES with PCM, it is worth highlighting the comparisons of different solar collectors with PCM integration within the collector itself. Figure 2.5 shows four geometries analyzed in the work. The design is to stabilize the temperatures of the collector during its operation as well as to have TES integrated in the solar collector.

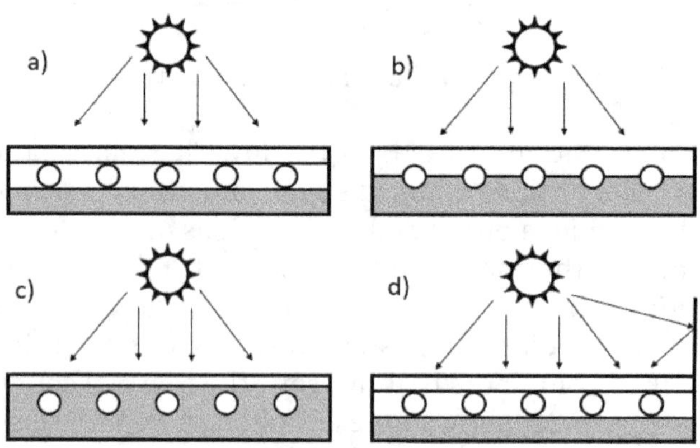

Figure 2.5 Schematic of a flat plate solar collector with PCM technologies (a) below tubes, (b) half perimeters of the tubes, (d) immersed tubes, (d) with reflector, (1) tubes, (2) absorber, (3) glass cover, (4) PCM, (5) air layer, (6) insulation and (7) reflector.

Recently low concentration systems are being developed for domestic and industrial applications. These collectors achieve higher fluid temperature. Some designs exhibit systems with PCM integration, as it can be seen in Figure 2.6. It was found the implementation of PCM in the solar collector tank extended the time for which it could supply hot water by up to 25% in 80 °C charging mode.

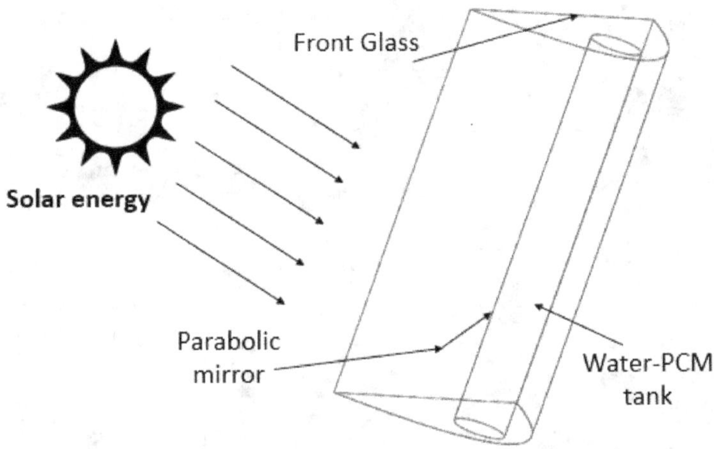

Figure 2.6 Cross-sectional schematic of the integrated collector storage solar water heater.

2.3.2 Concentrating Solar Power (CSP) Plants

Solar thermal power plants can generate large amounts of electricity through solar concentration systems that heat a fluid which is then expanded in a turbine. All these plants have parabolic mirror systems (linear or circular) which project the sun's rays onto the tube through which the fluid circulates or flat mirror systems that reflect the radiation in the focus of a tower. Unlike the low concentration systems, CSP plants have high concentration collectors, with concentrating ratios higher than 50. The higher the solar concentration is, the higher fluid temperature can be achieved and, therefore, the installation performance is improved. At present, there are four types of CPS technologies for the generation of electrical energy, which are schematically shown in Figure 2.7.

Due the high levels of concentration achieved in CSP, the temperature of the working fluid can be extremely high, over 500 °C. This feature of CSP allows high efficiency thermodynamic cycles to being able to generate a large amount of electrical energy.

Figure 2.7 Concentrating solar power installations: a) parabolic trough system, b) solar power tower, c) parabolic dish system and d) Fresnel reflectors (Source: NREL).

Currently, Spain is the country with the most operating CSP systems [37]. With 2304 MW installed, it is almost half of the power of the operational plants in the world. New CSPs will start operating in a few years, thus, the power capacity of this technology will be doubled. The most common technology is a parabolic troughs system [38], as is cheaper, simpler and easier to optimize as compared with the other technologies.

In the design of such facilities, special care is taken with the location with a large number of hours of sunshine per year, as the solar power plants require TES to generate electrical power without interruption for greater number of hours. In some installations, auxiliary boilers are installed, that allow generating electricity in periods of insufficient radiation.

Due to the high working temperature of these facilities and a large amount of energy that is necessary to store, the best alternative for TES is the use of PCM [39-42], with molten salts the most commonly used. Most of the operating plants currently use as PCM a mixture of 60% sodium nitrate and 40% potassium nitrate.

The installations with thermal accumulation have oversized collectors to generate an excess of energy that covers the turbine's energy demand for a period of between 4-8 hours [37,38]. This energy is stored in two large thermally insulated tanks (Figure 2.8), one cold and other hot. During daylight hours, the salts are pumped from the cold PCM tank to a heat exchanger, where the PCM increases its internal energy, and subsequently stored in the hot tank. When the solar radiation decreases or disappears, the cycle is reversed, where PCM is the medium that heats the working fluid of the installation in the exchanger to produce steam.

Figure 2.8 Salt tank for heat storage in CSP (Source: U.S. Department of Energy).

The geometry of the accumulation tanks is usually cylindrical and may contain tens of thousands of tonnes of PCM, depending on the number of hours of night operation and the power of the CSP plant. Figure 2.9 shows the outline of a complete installation of parabolic trough, where it is possible to see one of the most common distributions of heat storage tanks.

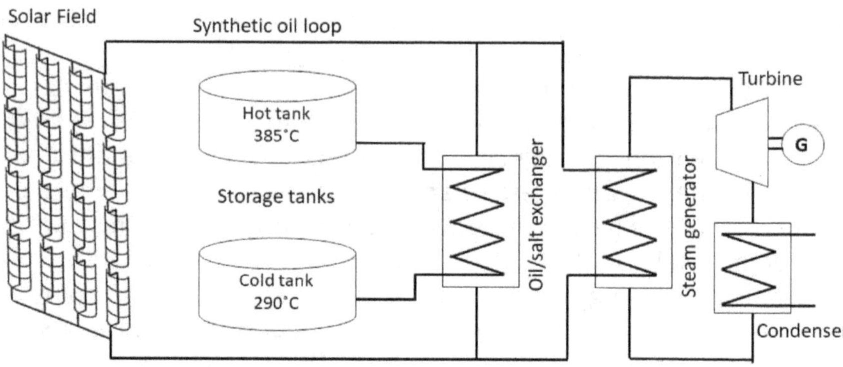

Figure 2.9 CSP installation diagram with thermal storage.

2.3.3 PCM in Photovoltaic (PV) and Hybrid Solar Energy (PV/T)

Efforts to improve the performance of solar energy are focused in two directions, improving the materials of the PV cells or cooling the PV panels. It is at this point where the PCMs are used as a tool to improve the performance of solar PV [43,44]. In order to maintain the efficiency of the PV cell, it is necessary to remove the solar energy absorbed that has not been transformed into electricity. This residual heat produces an overheating of the cells that causes a reduction in the efficiency of 0.4-0.65 %/K [45]. During operation, a PV panel can reach temperatures up to 80 °C. It is, therefore, necessary to keep the solar panel temperature as low as possible to increase the efficiency of the installation.

Adhering a volume of PCM on the back of the PV panel, it is possible to control the temperature of the panel [44]. PCM absorbs the residual heat in the form of latent heat, thus, stabilizing the panel temperature. Three of the first design concepts were tested and compared by Huang *et al.* [21,24]. The authors compared aluminum plate to simulate a PV cell, aluminum plate with a container filled with PCM to simulate a PV/PCM system and an aluminum plate with a container filled with PCM and integrated fins. Integrated PCM system with fins demonstrated effective temperature regulation of the plate. A schematic of the system tested by Huang *et al.* is presented in Figure 2.10, where temperatures below 34 °C were obtained. For optimal system operation, the PCM to be implemented in these cases must have a melting point between 25-40 °C and due to the low thermal conductivity of these materials, it is not recommended to work with thicknesses greater than 30 mm.

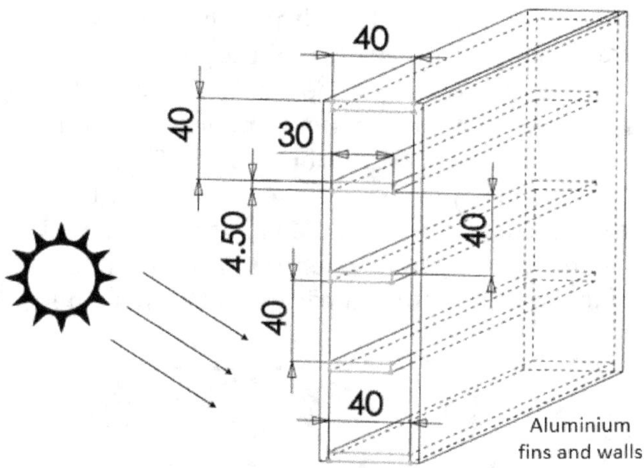

Figure 2.10 Outline and performance graph of PV panel with PCM.

The effects of convection and crystalline segregation in a PCM as a function of efficiency of heat transfer within the PV/PCM system has been experimentally investigated by Huang *et al.* [26]. The thermal performances of bulk PCM with crystallization segregation for different internal fin arrangements are presented. It is noted that the addition of internal fins improves the temperature control of the PV in a PV/PCM system. Two PCMs with different phase transient temperatures for improving the heat regulation have been predicted (see Figure 2.11) [26].

Hasan *et al.* [9] used five different PCM types (RT20, capric: palmitic acid, capric: lauric acid, calcium chloride and SP22) to enhance the thermal regulation of PV. Four different systems were tested under different insolation values (500 W/m^2, 750 W/m^2 and 1000 W/m^2) in a small-scale indoor experiment using a solar simulator. For insolation levels of 500 W/m^2, system A with capric: lauric and capric: palmitic was shown to maintain a lower PV temperature for the longest period (up to 2.5 hours at 10 °C lower than the reference system). At 750 W/m^2 and 1000 W/m^2, system A with calcium chloride was shown to maintain PV front surface temperatures 10 °C below the reference for a prolonged period of time compared to the other systems.

Also, a full-scale outdoor experiment was undertaken in Dublin,

Ireland and Vehara, Pakistan [9]. The PV panels were identical with 65 W capacity. One of the modules was used in the experiment as a reference and the other two were PV-PCM systems with eutectic mixture of capric-palmitic and salt hydrate. The results were compared at peak instantaneous temperature of the reference PV. Capric: palmitic PV-PCM system regulated the PV by 7 °C and the PV-PCM system with salt hydrate maintained a temperature reduction of 10 °C compared to the reference PV. Although in the PV-PCM configuration, the amount of electric power is higher compared to a standard PV panel, the heat absorbed by the PCM is not used. For this reason, hybrid PV/Thermal (PV/T) solar energy system was developed which converts solar radiation into electrical and thermal energy [25].

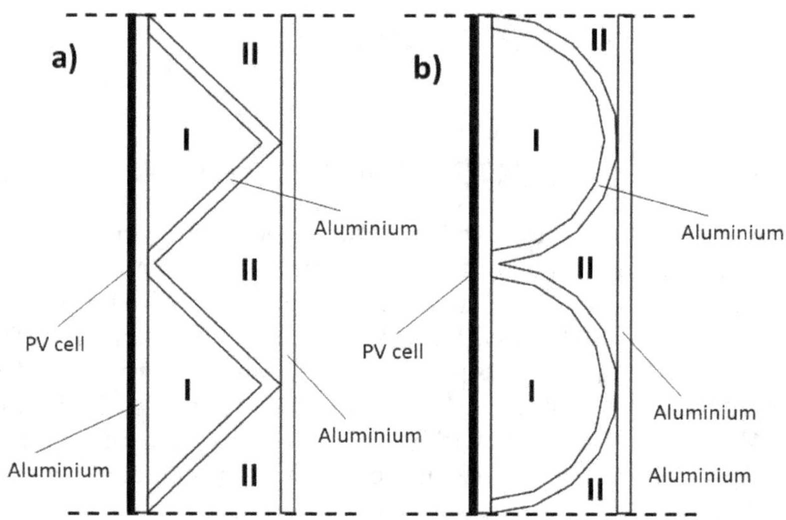

Figure 2.11 Schematic diagram of PV/PCM system with metal cells for different PCMs.

A PV/T (photovoltaic and thermal panel) system using phase change materials was designed and installed in Dublin to analyze its behavior and performance, as presented in Figure 2.12. The system was made from a standard PV panel which was disposed on a steel container full of PCM and equipped with piping network in the PCM. The aim of the set-up was to use the PCM to absorb the heat during the day and, therefore, make the PV panel more efficient by cooling it down and subsequently release the heat to the water piping in the evening and at night as the ambient temperature drops. Thus, the water going through the piping network would be heated at a more convenient time, such as when users come back home from work. This system was compared to three reference systems: a standard solar panel, a standard equipped with a steel container on the back of it and a PV equipped with a steel container and piping network inside the container (PV/T). Four used 80 W monocrystalline silicon solar panels were used in the experiment. The container was made from steel as it has a high thermal conductivity (16 W/m K) as well as a high resistance to corrosion which can be an issue while using phase change materials. ThePCM used for the PV/T/PCM system was capric: palmitic mixture (25% capric and 75% palmitic) which has a phase change between 16 and 25 °C.

Figure 2.12 Experimental set-up of PV-PCM systems, location of thermocouples and attachment of PV to PCM container. Adapted from Reference 25 with permission from Elsevier.

Adding the PCM improved the performance of PV/T system as water was heated at a higher temperature during a longer period. The maximum water temperature difference between the PV/T and PV/T/PCM reached 5.5 °C at 6:00 am. PV/T/PCM also showed to be more efficient than the other systems on the electrical side as the temperature of the system was regulated by PCM, thus, the panel was cooled down. The PCM improved the heat extraction up to seven times compared to the system without it [46].

Several review papers have been published on the use of PCM as a thermal management technique of PV, emphasizing the growing body of literature in the area [14,43,47,48]. Yin *et al.* [49] presented a PV/T system with integrated heat storage. Figure 2.13 presents a schematic design of different configurations of PV/T collectors for either air (right) or water (left).

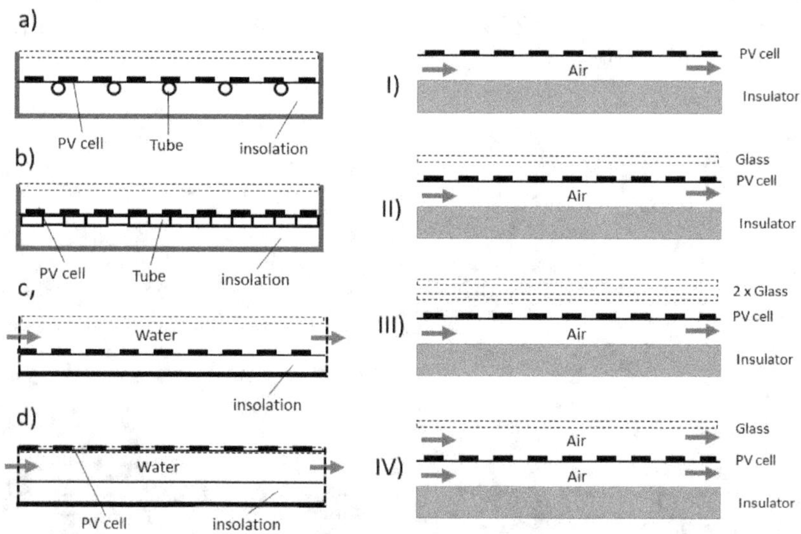

Figure 2.13 Diagram of collectors PV/T water (left) and air (right) [47].

The use of PCM in PV/T installations is in the heat accumulation tanks, just as in the case of conventional thermal solar panels. Several review papers have been published on the use of PCM as a thermal management technique of PV [14,43,49].

Subsequently, PV/T panels are being developed with modules that integrate a volume of PCM in the same PV/T panel. In this way, as in the PV-PCM, the temperature of the cells remains stable, in addition to the heat being stored in the PV/T panel itself. Figure 2.14 presents the design scheme proposed by Malvi *et al.* [50] for distribution of the PCM in the PV/T. The optimum thickness of PCM was found to be 30 mm which increases the PV efficiency by approximately 6.5%. The electrical output of the PV/T/PCM system was shown to increase by 9% when compared to a PV system only.

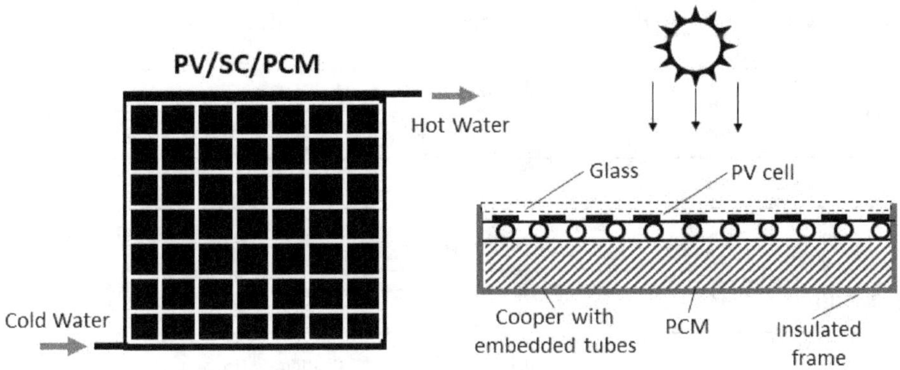

Figure 2.14 Schematic of a PV/T-PCM system.

Although it is possible to find numerous designs of systems in the literature to increase the efficiency of PV panels with PCM, most are not commercially exploited. The extra cost of including PV panel heat absorption systems and the complexity of some designs make it difficult for these technologies to be viable at present. It is only possible to find commercialized PV/T water systems [51] without integrating PCM.

2.4 Corrosion

Corrosion is a natural process which involves the deterioration of a material due the environment. In the case of the PCM in solar applications, it is critically important to study the corrosion between the PCM and the container material used for PCM encapsulation. As mentioned before, the most used PCMs in this field of applications are organic and salt inorganic PCMs, the corrosion behavior in each case is different.

2.4.1 Inorganic Compounds

Inorganic compounds are highly corrosive, for this reason it is necessary to perform the corrosion tests over different temperature ranges to ensure the container materials are compatible with their PCM. The most common inorganic PCM are the salts and salt hydrates. The most common salt hydrates for these applications are summarized in Table 2.1. This section also presents the compatibility of the salt hydrates with various materials.

Table 2.1 Summary of corrosion affects from inorganics PCMs

PCM	Mel. Pt. (°C)	Heat of fusion (kJ/kg)	Recommendations					Ref.
			Copper	Brass	Alumi-num	Stain less	Steel	
$ZnCl_2 \cdot 3H_2O$	10	---	Do not use	---	Do not use	Use	Do not use	52
$K_2HPO_4 \cdot 6H_2O$	13	---	Caution	---	Do not use	Use	Do not use	52
$NaOH \cdot 1.5H_2O$	15	---	Do not use	---	Do not use	Use	Caution	52
SP21E	21	160	Use	---	Caution	Use	Caution	53
$CaCl_2 \cdot 6H_2O$	36	170.5	Use	Use	Use	Use	Caution	54, 55, 61
TH29	29		Use	Use	Do not use	Cau-tion	Do not use	56
$Na_2SO_4 \cdot 10H_2O$	32	248	---	---	---	---	---	57
$Na_2HPO_4 \cdot 12H_2O$	35	280	Caution	Use	Do not use	Use	Caution	54, 55
$Zn(NO_3)_2 \cdot 6H_2O$	36	149.6	Do not use	Do not use	Do not use	Use	Do not use	54, 55
$Ca(NO_3)_2 \cdot 4H_2O$	42.7	---	Caution	---	---	---	Caution	58, 59
$K_3PO_4 \cdot 7H_2O$	45	---	Do not use	---	Do not use	Use	Use	52
$Zn(NO_3)_2 \cdot 4H_2O$	36	---	Do not use	---	Do not use	Do not use	Do not use	52
$Na_2S_2O_3 \cdot 5H_2O$	48	220	Do not use	Do not use	Caution	Use	Caution	56
$MgSO_4 \cdot 7H_2O$	48.5	202	Do not use	---	---	Use	Do not use	52
$NaOAc \cdot 3H_2O$	58	---	Caution	Caution	Use	Use	Use	56
$Mg(NO_3)_2 \cdot 6H_2O$	88.9	163	Caution	---	---	---	Caution	58, 59
$MgCl_2 \cdot 6H_2O$	114.5	135	Do not use	---	Do not use	Do not use	---	60

Zinc Chloride Trihydrate ($ZnCl_2 \cdot 3H_2O$)

Moreno *et al.* [52] carried out an exhaustive study of the behavior of the salt hydrated with different metals. Copper and stainless steel found non-significant corrosion after 12 weeks, where at four weeks there was a corrosion rate of 4.4 mg/cm^{2}yr and at 12 weeks 1.7 mg/cm^{2}yr in the case of copper, however, the corrosion rate was negligible in the case of stainless steel. Carbon steel was highly corroded after 12 weeks, also a decrease of the corrosion rate with time was detected. Aluminum was the most affected with aggressive corrosion after 1 week.

Potassium Hydrogen Phosphate Hexahydrate ($K_2HPO_4 \cdot 6H_2O$)

This salt hydrate was studied by Moreno *et al.* [52] with copper and stainless steel. The authors found non-significant corrosion, however, a high corrosion rate at the beginning was detected with carbon steel and aluminum, however, after 4 weeks, the corrosion rate decreased. It continued to decrease in the case of carbon steel after 12 weeks, but increased again with aluminum.

Sodium Hydroxide Hydrated ($NaOH \cdot 1.5H_2O$)

The behavior of $NaOH \cdot 1.5H_2O$ with copper showed some corrosion, but the sample did not present any change. With stainless steel, the corrosion was observed to be negligible. Carbon steel initially had a high corrosion rate than after four and twelve weeks. Very aggressive behavior was observed with aluminum [52].

Calcium Chloride Hexahydrate ($CaCl_2 \cdot 6H_2O$)

Ren *et al.* [61] studied the behavior of the salt hydrated with two types of aluminum, carbon steel and copper at different temperatures (30, 50 and 80 °C). The authors found that the corrosion of copper increased with temperature and time, where the solution became blue at 50 °C due the corrosive products and a vivid blue solution was detected at 80 °C. Moreover, no pitting corrosion was detected. In the case of carbon steel, the mass loss increased with time and the corrosion rate decreased. The authors detected an oxide layer on the sample that could act as a protection layer. Also, the solution was transparent-yellow and dark precipitates were found at 50 and 80 °C, along with no detection of pitting corrosion. In aluminum 5086, the mass loss increased and the corrosion rate decreased with time due to the formation of a protective oxide layer. At 30 °C, there was no pitting corrosion detected. However, at 50 °C and 80 °C, minimal pitting areas was found. Aluminum 6061 presented small pitting areas at 30 and 50 °C after 14 weeks, but pitting corrosion appeared at 80 °C after four weeks and increased with time.

Cabeza *et al.* [54,55] found that aluminum immersed in this salt was covered with a non-continuous layer after three days. After two weeks, the solution pH changed from 6 to 7/8 as a consequence of the $Al(OH)_3$ products in the solution. After 72 days, the sample was partly covered with a black layer. Copper, brass and stainless steel presented no change in the solution or in the sample and the results of mass loss and corrosion rate were negligible. However, the solution became yellow with steel and the sample was covered with a black layer after three days. The solution was bright yellow with brown precipitates after two weeks. The mass loss increased after two weeks and the corrosion rate decreased with time.

Sodium Sulfate Decahydrate ($Na_2SO_4 \cdot 10H_2O$), Glaubers Salt

García-Romero *et al.* [57] studied the behavior of this salt in contact with different aluminum types. The test was carried out with samples partly and fully immersed in the PCM. The 2024 and 1050 alloy showed increased corrosion after 30 days, but mass loss was negligible when partly immersed. However, fully immersed material remained non-corroded, which confirmed the corrosion mechanism observed in the aerated specimens, due to different oxygen concentration in different zones. In the case of 3003 and 6063 alloys, no sign of corrosion and no loss of brightness in any case (full or partly immersed) were observed.

Sodium Hydrogen Phosphate Dodecahydrate ($Na_2HPO_4 \cdot 12H_2O$)

Cabeza *et al.* [54,55] found the aluminum in contact with this salt hydrate presented significant extent of white precipitate ($Al(OH)_3$) after three days with a mass loss of 35 mg. The corrosion rate was high after three days and subsequently decreased with time. In contact with brass, copper and stainless steel, no change in the samples was observed and the corrosion results were statistically insignificant.

Zinc Nitrate Hexahydrate ($Zn(NO_3)_2 \cdot 6H_2O$)

Cabeza *et al.* [54,55] studied this PCM in short and medium term and in contact with five different materials: aluminum, brass, copper, stainless steel and steel. In the case of aluminum, the mass loss increased with the time, but no increase of mass loss was detected after two weeks. The sample presented a layer of white precipitate and the solution had the same precipitate, which was due to the presence of hydroxide of aluminum ($Al(OH)_3$). The formation of this layer led to a decrease of the corrosion rate. After 75 days, the samples were highly corroded and the presence of the white precipitate increased. The behavior with brass was different, the solution became blue and brown precipitates were detected. The mass loss increased with time and the corrosion rate decreased, however, was higher than the aluminum samples. With copper, a blue solution and brown precipitates were also seen. The sample was fully covered by a grey layer after three days, which became black after one week. The mass loss at three days and one week was the same as brass, which did not increase subsequently, while the corrosion rate was higher after three days, followed by a decrease. After 75 days, the sample was severed with a marked increase in mass loss, and the solution became yellow with blue precipitate. Stainless steel did not show any change, but the solution became yellow as a product of the dissolution of iron ions (Fe^{3+}). The mass loss and the corrosion rate were negligible. After 75 days, the sample showed nearly no corrosion, however, brown precipitate was visible. In the case of steel, brown precipitate was detected after three days, which increased with time. The weight loss was significant, and the corrosion rate was the highest.

Calcium Nitrate Tetrahydrate ($Ca(NO_3)_2{\cdot}4H_2O$)

The compatibility of this salt was studied with carbon steel. After 700 hours, mass loss and an insoluble layer of 30 μm were observed, followed by the detection of cracks in the layer [59]. With copper, a non-compact layer of Cu_2O with precipitation products of $Cu_2(NO_3)(OH)_3$ were observed. Up until 300 hours, mass loss was detected, however, the formation of a layer was observed afterwards [58].

Tripotassium Phosphate Heptahydrate ($K_3PO_4{\cdot}7H_2O$)

The corrosion behavior of this salt hydrate with copper was initially higher, but still had lower values after four and twelve weeks in contrast with aluminum which underwent aggressive corrosion under same conditions. Stainless steel and carbon steel showed very low rates of corrosion [52].

Zinc Nitrate Tetrahydrate ($Zn(NO_3)_2{\cdot}4H_2O$)

Zinc nitrate tetrahydrate showed very high corrosion rate in contact with copper (1 week 1167 mg/cm^{2}yr, 4 weeks 644 mg/cm^{2}yr and 12 weeks 338.5 mg/cm^{2}yr), carbon steel (1 week 1741 mg/cm^{2}yr, 4 weeks 484.4 mg/cm^{2}yr and 12 weeks 196.2 mg/cm^{2}yr) and aluminum (1 week 181.5 mg/cm^{2}yr, 4 weeks 206.2 mg/cm^{2}yr and 12 weeks 162.2 mg/cm^{2}yr). However, the corrosion rate was negligible with stainless steel [52].

Sodium Thiosulfate Pentahydrate ($Na_2S_2O_3{\cdot}5H_2O$)

Moreno *et al.* [52] studied the behavior of this salt with common materials. Copper showed a very high corrosion rate after one week with a maximum of 2663.5 mg/cm^{2}yr, which subsequently decreased to 437 mg/cm^{2}yr and remained constant until 12 weeks. Carbon steel and aluminum had a similar result, a value of 40 mg/cm^{2}yr after 1 week, 15 mg/cm^{2}yr after 4 weeks and at 5 mg/cm^{2}yr after 12 weeks. Stainless steel did not show any signs of corrosion and the results were negligible.

Cabeza *et al.* [56] found that aluminum and stainless steel were corrosion resistant with this salt hydrate. However, brass and copper had a significant mass loss and both had a black precipitate after few days. Steel was seen to have corrosion with insignificant mass loss, but black precipitate appeared after 70 days.

Magnesium Sulfate Heptahydrate ($MgSO_4{\cdot}7H_2O$)

This salt hydrate showed a corrosive behavior initially in contact with copper with a value of 121 mg/cm^{2}yr, but dropped to 43 mg/cm^{2}yr after four weeks. After twelve weeks, it dropped again to 21 mg/cm^{2}yr. In contact with carbon steel, a very high value of 749.8 mg/cm^{2}yr after one week was observed, which decreased to 277.4 mg/cm^{2}yr at four weeks and ended with a value of 80.5 mg/cm^{2}yr at twelve weeks. In contrast, aluminum and stainless steel had low corrosion rates [52].

Sodium Acetate Trihydrate ($NaOAc{\cdot}3H_2O$)

The effect of sodium acetate trihydrate with common materials was studied by Cabeza et al. [56]. The authors detected that aluminum, steel and stainless steel did not present corrosion and the solution did not exhibit any change. In the case of brass and copper, mild corrosion was detected after 70 days.

Magnesium Nitrate Hexahydrate ($Mg(NO_3)_2{\cdot}6H_2O$)

In the case of carbon steel, it was found to have a weight decrease during the first 500 hours. Afterwards, a constant corrosion rate of 11.9 mg/cm^{2}yr was maintained. Copper exhibited a different reaction, where an increased mass loss was found at the beginning due to the formation of oxide layer (Cu_2O) and oxide products of $Cu_2(NO_3)(OH)_2$, with a corrosion rate of 8.6 mg/cm^{2}yr [58]. With aluminum, during the first 500 hours of the test, the mass loss was constant and subsequently increased markedly with a corrosion rate of 2.9 mg/cm^{2}yr [54].

2.4.2 Organic Compounds

Organic compounds are recommended for use in contact with metals, because they are less reactive. Some studies show that a corrosive behavior is detected in some cases. Table 2.2 summarizes the properties and the compatibility of different PCMs with common materials and their behavior.

Capric Acid

Browne *et al.* [62] showed that capric acid had aggressive corrosive behavior with different materials. The authors showed that copper and brass were affected by pitting corrosion, but the corrosion rate was higher in the case of copper than brass. Aluminum was also affected by pitting corrosion after 540 days and the corrosion got progressively worse with a mass loss of 6% at the end of the test. Mild steel was adversely affected by this PCM, however no deep pitting was detected. In the case of stainless steel and Perspex, no sign of corrosion was detected, and the results of the corrosion tests were negligible.

Capric Acid (73.5%) + Myristic Acid (26.5%)

Ferrer *et al.* [53] detected that copper underwent corrosion and the solution became blue. In the case of carbon steel, a high extent of corrosion was observed in the first week which may be explained by the passivation of the sample. No corrosion was detected at four and twelve weeks. Stainless steel 304, 316 and aluminum exhibited relatively low corrosion rate, under 1 mg/cm^2yr.

Table 2.2 Summary of corrosion effects using organic PCMs

PCM	Melting point (°C)	Heat fusion (kJ/kg)	Recommendations					Ref.
			Copper	Brass	Aluminum	Stainless Steel	Carbon Steel	
Capric acid	27.5-32.7	167	Caution	Caution	Caution	Use	Caution	62
Capric acid (73.5%) + myristic acid (26.5%)	21.4	152	Caution	---	Use	Use	Use	53
Capric acid (75.2%) + palmitic acid (24.8%)	17.7-22.8	153	Caution	Caution	Caution	Use	Caution	53,62
Capric acid (75.2%) + Lauric acid (24.8%)	18.9-22.3	116	Caution	Caution	Caution	Use	Caution	62
Lauric acid	42.6	211.6	Caution	---	Use	Use	Use	63
Myristic acid	53.8	192.0	Caution	---	Use	Use	Caution	63
Stearic Acid	53.8	174.6	Caution	---	Use	Use	Use	63
Palmitic acid	59.8	197.9	Caution	---	Use	Use	Caution	63
Glutaric acid			Caution	---	Caution	Use	---	64

Capric Acid (75.2%) + Palmitic Acid (24.8%)

Ferrer *et al.* [53] studied the behavior of this eutectic fatty acid PCM in contact with common materials. The authors detected that the only material that experienced a remarkable weight loss during the 12 weeks of test was copper. Materials such as carbon steel exhibited passivation at the beginning due to the mass loss at the first week and subsequent constant value.

Lauric Acid

The behavior of lauric acid with common materials was tested under 910 thermal cycles by Sari and Kaygusuz [63]. The authors observed that stainless steel (SS 304 L), carbon steel (C 20) and aluminum did not exhibit any sign of corrosion on the surface, however, copper was slightly corroded. In terms of corrosion values, copper had the maximum values with a 15.232 mg/cm^2 of mass loss and $1.7 \cdot 10^2$ mg/day corrosion rate.

Myristic Acid

Myristic acid also had a corrosive behavior under 910 thermal cycles. Slight corrosion on the surface of carbon steel (C20) and copper was detected, with corrosion values (mass loss and corrosion rate) of 44.620 mg/cm^2 and $4.9 \cdot 10^2$ mg/day in the case of carbon steel and 15.437 mg/cm^2 and $1.6 \cdot 10^2$ mg/day with the copper. In contrast, stainless steel (SS 304 L) and aluminum did not show any sign of corrosion and the corrosion values were observed to be negligible for these materials [63].

Stearic Acid

Sari and Kaygusuz [63] observed that stainless steel (SS 304 L), carbon steel (C 20) and aluminum did not have any sign of corrosion on the surface, however, copper was mildly corroded with corrosion values (mass loss and corrosion rate) of 31.416 mg/cm^2 and $3.4 \cdot 10^2$ mg/day, under 910 thermal cycles.

Palmitic Acid

Stainless steel and aluminum were found to be corrosion resistant in contact with palmitic acid under 910 thermal cycles with low corrosion values. In contrast, copper and carbon steel (C20) were found to have slight corrosion on the surface [63].

Glutaric Acid

Raam Dheep and Sreekumar [64] observed that aluminum in contact with glutaric acid had a total mass loss of 11.2% with a corrosion rate of 10.9869 mg/cm²yr. Also, pits on the aluminum surface were observed. In the case of copper, the total mass loss was 6.5% with a corrosion rate of 0.9423 mg/cm²yr and the solution became green after sixty thermal cycles which occurred due to oxidation of the sample. Large pits were detected on the surface in this case. Stainless steel had better results with a mass loss of 1.41% and corrosion rate of 0.3004 mg/cm²yr, but small pits were detected due to the rupture of the passive layer.

2.5 Conclusion

Solar energy has a wide variety of applications in both domestic and industrial use. This energy allows us to obtain, directly and without GHG emissions, heat and electricity at a cost that is currently competitive with other more polluting technologies. However, the availability of this energy is limited to daylight hours, thus, requiring efficient thermal storage systems.

The use of PCMs in solar energy technologies has allowed the option of storing thermal energy produced by the solar collectors during the day so that it is possible to have energy available during the night.

In addition, the use of PCMs in PV energy also allows increasing the efficiency of the PV modules by increasing the electricity generation. An innovative design of PV energy is hybrid solar energy (PV/T), where the overall panel efficiency is increased thanks to the use of the thermal energy generated in the PV module. The PCM volume provides a built-in heat accumulation system inside the PV/T module, saving space and simplifying the final installation. The choice of PCM used in each solar energy application must be made considering both the thermophysical properties of the PCM and the compatibility between the different materials of the system, as has been outlined in this chapter.

Acknowledgements

David González Peña thank Junta de Castilla-León for economic support (PIRTU Program, ORDEN EDU/310/2015).

References

1. Atkin, P., and Farid, M. M. (2015) Improving the efficiency of photovoltaic cells using PCM infused graphite and aluminium fins. *Solar Energy*, **114**, 217-228.
2. Evans, A., Strezov, V., and Evans, T. J. (2009) Assessment of sustainability indicators for renewable energy technologies. *Renewable and Sustainable Energy Reviews*, **13**(5), 1082-1088.
3. Renewable Power Generation Costs in 2017, *IRENA* (2018). Online: https://www.irena.org/publications/2018/Jan/Renewable-power-generation-costs-in-2017 [accessed 19th December 2018].
4. Biyik, E., Araz, M., Hepbasli, A., Shahrestani, M., Yao, R., Shao, L., Essah, E., Oliviera, A. C., del Cano, T., Rico, E., Lechon, J. L., Andrade, L., Mendes, A., and Atli, Y. B. (2017) A key review of building integrated photovoltaic (BIPV) systems. *Engineering Science and Technology, an International Journal*, **20**(3), 833-858.
5. Oliver, M., and Jackson, T. (2001) Energy and economic evaluation of building-integrated photovoltaics. *Energy*, **26**(4), 431-439.
6. Collares-Pereira. M., Goodman, N. B., Greenman, P., O'Gallagher, J., Rabl, A., Simmons, H., Wharton, L., and Winston, R. (1978) Compound parabolic concentrators with non-evacuated receivers - Prototype performance and a larger scale demonstration in a school heating system. *Sun: Mankind's Future Source Energy*, **2**, 1227-1232.
7. Zacharopoulos, A., Eames, P. C., McLarnon, D., and Norton, B. (2000) Linear dielectric non-imaging concentrating covers for PV integrated building facades. *Solar Energy*, **68**, 439-452.
8. Stropnik, R., and Stritih, U. (2016) Increasing the efficiency of PV panel with the use of PCM. *Renewable Energy*, **97**, 671-679.
9. Hasan, A., McCormack, S. J., Huang, M. J., and Norton, B. (2010) Evaluation of phase change materials for thermal regulation enhancement of building integrated photovoltaics. *Solar Energy*, **84**, 1601-1612.
10. Browne, M. C., Norton, B., and McCormack, S. J. (2016) Heat retention of a photovoltaic/thermal collector with PCM. *Solar Energy*, **133**, 533-548.
11. Ames, W. (1983) Photovoltaic Devices for Producing Electrical and Heat Energy, patent US 4389533.
12. Pal, D., and Joshi, Y. K. (1999) Thermal Control of Horizontally Mounted Heat Sources using Phase Change Materials. *Proceedings of InterPACk'99*, volume 2, pp. 1625-1630.
13. Griffiths, P. W., and Eames, P. C. (2007) Performance of chilled ceiling panels using phase change material slurries as the heat transport medium. *Applied Thermal Engineering*, **27**(10), 1756-1760.
14. Browne, M. C., Norton, B., and McCormack, S. J. (2015) Phase change

materials for photovoltaic thermal management. *Renewable and Sustainable Energy Reviews*, **47**(244), 762-782.

15. Gunther, E., Hiebler, S., Mehling, H., and Redlich, R. (2009) Enthalpy of phase change materials as a function of temperature: Required accuracy and suitable measurement methods. *International Journal of Thermophysics*, **30**(4), 1257-1269.
16. Cabeza, L. F., Castell, A., Barreneche, C., De Gracia, A., and Fernández, A. I. (2011) Materials used as PCM in thermal energy storage in buildings: A review. *Renewable and Sustainable Energy Reviews*, **15**(3), 1675-1695.
17. Pereira da Cunha, J., and Eames, P. (2016) Thermal energy storage for low and medium temperature applications using phase change materials - A review. *Applied Energy*, **177**, 227-238.
18. Xu, B., Li, P., and Chan, C. (2015) Application of phase change materials for thermal energy storage in concentrated solar thermal power plants: A review to recent developments. *Applied Energy*, **160**, 286-307.
19. Siegel, R. (1977) Solidification of low conductivity material containing dispersed high conductivity particles. *International Journal of Heat and Mass Transfer*, **20**(10), 1087-1089.
20. Cabeza, L. F., Mehling, H., Hiebler, S., and Ziegler, F. (2002) Heat transfer enhancement in water when used as PCM in thermal energy storage. *Applied Thermal Engineering*, **22**(10), 1141-1151.
21. Huang, M. J., Eames, P. C., and Norton, B. (2004) Thermal regulation of building-integrated photovoltaics using phase change materials. *International Journal of Heat and Mass Transfe*r, **47**(12-13), 2715-2733.
22. Huang, M. J. Eames, P. C., and Hewitt, N. J. (2006) The application of a validated numerical model to predict the energy conservation potential of using phase change materials in the fabric of a building. *Solar Energy Materials and Solar Cells*, **90**(13), 1951-1960.
23. Hasan, A., McCormack, S., Huang, M., and Norton, B. (2014) Energy and cost saving of a photovoltaic-phase change materials (PV-PCM) system through temperature regulation and performance enhancement of photovoltaics. *Energies*, **7**(3), 1318-1331.
24. Huang, M. J., Eames, P. C., and Norton, B. (2006) Phase change materials for limiting temperature rise in building integrated photovoltaics. *Solar Energy*, **80**(9), 1121-1130.
25. Browne, M. C., Lawlor, K., Kelly, A., Norton, B., and Cormack, S. J. (2015) Indoor characterisation of a photovoltaic/thermal phase change material system. *Energy Procedia*, **70**, 163-171.
26. Huang, M. J., Eames, P. C., Norton, B., and Hewitt, N. J. (2011) Natural convection in an internally finned phase change material heat sink for the thermal management of photovoltaics. *Solar Energy Materials and Solar Cells*, **95**(7), 1598-1603.

27. Guney, M. S. (2016) Solar power and application methods. *Renewable and Sustainable Energy Reviews*, **57**, 776-785.
28. Mekhilef, S., Saidur, R., and Safari, A. (2011) A review on solar energy use in industries. *Renewable and Sustainable Energy Reviews*, **15**(4), 1777-1790.
29. Kalogirou, S. A. (2004) Solar thermal collectors and applications. *Progress in Energy and Combustion Science*, **30**(3), 231-295.
30. Sharif, M. K. A., Al-Abidi, A. A., Mat, S., Sopian, K., Ruslan, M. H., Sulaiman, M. Y., and Rosli, M. A. M. (2015) Review of the application of phase change material for heating and domestic hot water systems. *Renewable and Sustainable Energy Reviews*, **42**, 557-568.
31. Kenisarin, M., and Mahkamov, K. (2007) Solar energy storage using phase change materials. *Renewable and Sustainable Energy Reviews*, **11**(9), 1913-1965.
32. Mazman, M., Cabeza, L. F., Mehling, H., Nogues, M., Evliya, H., and Paksoy, H. O. (2009) Utilization of phase change materials in solar domestic hot water systems. *Renewable Energy*, **34**(6), 1639-1643.
33. Padovan, R., and Manzan, M. (2014) Genetic optimization of a PCM enhanced storage tank for solar domestic hot water systems. *Solar Energy*, **103**, 563-573.
34. Nallusamy, N., Sampath, S., and Velraj, R. (2007) Experimental investigation on a combined sensible and latent heat storage system integrated with constant/varying (solar) heat sources. *Renewable Energy*, **32**(7), 1206-1227.
35. Ibanez, M., Cabeza, L. F., Sole, C., Roca, J., and Nogues, M. (2006) Modelization of a water tank including a PCM module. *Applied Thermal Engineering*, **26**(11-12), 1328-1333.
36. Wang, Z. Qiu, F. Yang, W. and Zhao, X. (2015) Applications of solar water heating system with phase change material. *Renewable and Sustainable Energy Reviews*, **52**, 645-652.
37. Tian, Y., and Zhao, C. Y. (2013) A review of solar collectors and thermal energy storage in solar thermal applications. *Applied Energy*, **104**, 538-553.
38. *Concentrating Solar Power Projects*, NREL, USA. Online: https://solarpaces.nrel.gov [accessed 17th December 2018].
39. Alva, G., Liu, L., Huang, X., and Fang, G. (2017) Thermal energy storage materials and systems for solar energy applications. *Renewable and Sustainable Energy Reviews*, **68**, 693-706.
40. Bauer, T., Pfleger, N., Laing, D., Steinmann, W. D., Eck, M., and Kaesche, S. (2013) High-temperature molten salts for solar power application. In: *Molten Salts Chemistry*, Lantelme, F., and Groult, H. (eds.), Elsevierm, USA, pp. 415-438.
41. Nomura, T., Okinaka, N., and Akiyama, T. (2010) Technology of latent heat storage for high temperature application: A review. *ISIJ International*, **50**(9), 1229-1239.

42. Sarvghad, M., Maher, S. D., Collard, D., Tassan, M., Will, G., and Steinberg, T. A. (2018) Materials compatibility for the next generation of concentrated solar power plants. *Energy Storage Material*, **14**, 179-198.
43. Ma, T., Yang, H., Zhang, Y., Lu, L., and Wang, X. (2015) Using phase change materials in photovoltaic systems for thermal regulation and electrical efficiency improvement: A review and outlook. *Renewable and Sustainable Energy Reviews*, **43**, 1273-1284.
44. Ling, Z. Zhang, Z. Shi, G., Fang, X., Wang, L., Gao, X., Fang, Y., Xu, T., Wang, S., and Liu X. (2014) Review on thermal management systems using phase change materials for electronic components, Li-ion batteries and photovoltaic modules. *Renewable and Sustainable Energy Reviews*, **31**, 427-438.
45. Radziemska, E. (2003) The effect of temperature on the power drop in crystalline silicon solar cells. *Renewable Energy*, **28**(1), 1-12.
46. Browne M. C., Quigley, D., Hard, H. R., Gilligan, S., Ribeiro, N. C. C., Almeida, N., and McCormack, S. J. (2016) Assessing the thermal performance of phase change material in a photovoltaic/thermal system. *Energy Procedia*, **91**, 113-121.
47. Makki, A., Omer, S., and Sabir, H. (2015) Advancements in hybrid photovoltaic systems for enhanced solar cells performance. *Renewable and Sustainable Energy Reviews*, **41**, 658-684.
48. Al-Waeli, A. H. A., Sopian, K., Kazem, H. A., and Chaichan, M. T. (2017) Photovoltaic/Thermal (PV/T) systems: Status and future prospects. *Renewable and Sustainable Energy Reviews*, **77**, 109-130.
49. Yin, H. M., Yang, D. J., Kelly, G., and Garant, J. (2013) Design and performance of a novel building integrated PV/thermal system for energy efficiency of buildings. *Solar Energy*, **87**, 84-195.
50. Malvi, C. S., Dixon-Hardy, D. W., and Crook, R. (2011) Energy balance model of combined photovoltaic solar-thermal system incorporating phase change material. *Solar Energy*, **85**(7), 1440-1446.
51. *Hybrid*, Solimpeks Solar Corp. (2016). Online: http://www.solimpeks.com/tax-product/hybrid/ [accessed 07th January 2019].
52. Moreno, P., Miro, L., Sole, A., Barreneche, C., Sole, C., Martorell, I., and Cabeza, L. F. (2014) Corrosion of metal and metal alloy containers in contact with phase change materials (PCM) for potential heating and cooling applications. *Applied Energy*, **125**, 238-245.
53. Ferrer, G., Sole, A., Barreneche, C., Martorell, I., and Cabeza, L. F. (2015) Corrosion of metal containers for use in PCM energy storage. *Renewable Energy*, **76**, 465-469.
54. Cabeza, L. F., Illa, J., Roca, J., Badia, F., Mehling, H., Hiebler, S., and Ziegler, F. (2001) Middle term immersion corrosion tests on metalsalt hydrate pairs used for latent heat storage in the 32 to 36 °C temperature range. *Materials and Corrosion*, **52**(10), 748-754.
55. Taylor, M. M., Cabeza, L. F., Marmer, W. N., and Brown, E. M. (2001)

Enzymatic modification of hydrolysis products from collagen using a microbial transglutaminase. I. Physical properties. *Journal of American Leather Chemists Association*, **96**(9), 319-332.

56. Cabeza, L. F., Roca, J., As, M. N., Mehling, H., and Hiebler, S. (2002) Immersion corrosion tests on metal-salt hydrate pairs used for latent heat storage in the 48 to 58°C temperature range. *Materials and Corrosions*, **53**, 902-907.
57. Garcia-Romero, A., Delgado, A., Urresti, A., Martin, K., and Sala, J. M. (2009) Corrosion behaviour of several aluminium alloys in contact with a thermal storage phase change material based on Glauber's salt. *Corrosion Science*, **51**(6), 1263-1272.
58. Danielik, V., Soska, P., and Felgerova, K. (2016) Corrosive effects of nitrate-containing phase change materials used with copper. *Acta Chimica Slovaca*, **9**(2), 75-83.
59. Danielik, V., Soska, P., Felgerovi, K., and Zemanovi, M. (2017) The corrosion of carbon steel in nitrate hydrates used as phase change materials. *Materials and Corrosions*, **68**(4), 416-422.
60. Gutierrez, A., Ushaka, S., Mamania, V., Vargas, P., Barreneche, C., Cabeza, L. F., and Grageda M. (2017) Characterization of wastes based on inorganic double salt hydrates as potential thermal energy storage materials. *Solar Energy Materials and Solar Cells*, **170**, 149-159.
61. Ren, S. J., Charles, J., Wang, X. C., Nie, F. X., Romero, C., Neti, S., Zheng, Y., Hoenig, S., Chen C., Cao F., Bonner, R., and Pearlman, H. (2017) Corrosion testing of metals in contact with calcium chloride hexahydrate used for thermal energy storage. *Materials and Corrosion*, **68**(10), 1046-1056.
62. Browne, M. C., Boyd, E., and Mccormack, S. J. (2017) Investigation of the corrosive properties of phase change materials in contact with metals and plastic. *Renewable Energy*, **108**, 555-568.
63. Sari, A., and Kaygusuz, K. (2003) Some fatty acid used for latent heat storage: thermal stability and corrosion of metals with respect to thermal cycling. *Renewable Energy*, **28**, 939-948.
64. Raam Dheep G., and Sreekumar, A. (2018) Investigation on thermal reliability and corrosion characteristics of glutaric acid as an organic phase change material for solar thermal energy storage applications. *Applied Thermal Engineering*, **129**, 1189-1196.

3

A Brief Introduction and Review of Shape Stabilized Composite Phase Change Materials (SSCPCM)

Xu Qiao, Xiangfei Kong,* Lu Wang and Chang Xi

School of Energy and Environmental Engineering, Hebei University of Technology, Tianjin 300401, China

**Corresponding author*: xfkong@hebut.edu.cn

3.1 Background

With the continuous development of our society, a rising demand of various resources has resulted in ever-increasing resource shortage and inefficient energy usage. Thus, seeking a renewable energy solution and an efficient way of energy consumption have become unprecedentedly important. Currently, fossil energy based on oil, natural gas and coal is still widely used and regarded as the dominant energy resource across the world. Fortunately, an increasing number of renewable energy technologies are continuously introduced and promoted in our daily life such as wind energy, solar energy, nuclear energy, etc. In this respect, scientists are making unremitting efforts to alleviate the looming energy crisis. As human society enters the new century, sustainable development has gradually become the ultimate focus. Specifically, environmental and energy issues have gradually become a topic of constant concern, especially the built environment on which human beings depend. The living and working environment are closely related to the building environment, thus, inseparable from people's daily lives. The rapid population growth and the increasing demand for living and working environment as well as building functions have added to the energy and environmental crisis we face. According to IEA statistics, in the past two decades (1984~2004), the world's primary energy consumption increased by 49%. In developed countries, civil and commercial building energy consumption accounted for between 20~40% of total social energy consumption, while other type of buildings having higher and

Phase Change Materials, edited by Vikas Mittal

steadily rising energy consumption [1]. As for developing countries, this figure is relatively higher. From the perspective of building energy consumption, we can develop significant energy-saving potential.

In recent years, building energy conservation has been paid significant attention. In building energy consumption, heating and air conditioning systems account for a considerable part of the total energy consumption. In the United States, for example, the energy consumption of the air conditioning system accounts for 50% of the total energy consumption of the buildings and 20% of the national energy consumption [1]. In today's rapidly developing society, people are increasingly demanding for the quality of indoor working and living environments. As we all know, the indoor environmental comfort has a very important impact on the health of the occupants or users and the production efficiency. The indoor environment is mainly regulated by heating, ventilation and air-conditioning system (HVAC system). The adjustment is mainly the indoor temperature and environmental humidity of the buildings. Thus, with today's trend increasingly focused on sustainable development, we must focus on how to reduce the energy consumption of HVAC systems. The main factors affecting the indoor thermal and humidity environment are sunlight, outdoor temperature, indoor equipment operation and human work intensity. This chapter mainly summarizes the effect of outdoor temperature on composite phase change materials (CPCMs) for reducing the influence on indoor environment as well as the combination of CPCMs and building materials to form shape stabilized composite phase change material wallboards (SSCPCMWs). The main aspects studied in the chapter are classification of CPCMs (including support materials and PCMs), enhancement of thermal conductivity of CPCMs, fabrication methods and properties of CPCMs. Starting with the brief description of PCMs and SSCPCMWs used for building envelopes, the classification and properties of PCMs are summarized, so is the development of PCMs.

3.2 Composite Phase Change Materials (CPCMs)

3.2.1 An Overview of PCMs

It is a phase change process that transforms a material from one phase to another. The physical and chemical properties of the material system are identical, and the uniform part with obvious interface with other parts is called a phase. Corresponding to the three states of solid, liquid and gas, the substance has a solid phase, a liquid phase and a gaseous phase. In the case of a phase change, there is a change in volume with the absorption or release of heat. This type of phase change is called "first-order phase change". For example, at atmospheric pressure and 0 °C temperature, 1 kg of ice is converted into water at the same temperature, and 79.6 kcal of heat is absorbed. At the same time, the volume also shrinks. Therefore, the transition between ice and water is a first-order phase change.

If the volume does not change during the phase change process, and the absorption and release of heat are not accompanied, but the physical quantities such as heat capacity, thermal expansion coefficient and isothermal compression coefficient change, such change is called a secondary phase transition. The transition between normal liquid helium (helium I) and superfluid helium (helium II), the transition between normal conductor and superconductor, the transition between paramagnetic and ferromagnetic, the transition between the ordered and disordered states of the alloy are examples of typical secondary phase transitions.

PCMs are quite common in our daily lives. Water is the most commonly seen PCM. As a new material in the construction industry, PCM refers to a substance that changes the state of matter with temperature and provides latent heat. The process of transforming physical properties is called a phase change process, in which case the PCM will absorb or release a large amount of latent heat. PCMs have the potential of becoming the best green carrier for energy conservation and environmental protection. PCMs have been listed as a national R&D utilization priority in China.

PCMs can be classified into organic and inorganic PCMs. These can also be divided into hydrated salt PCMs and waxy PCMs. The most common PCM is water. When the temperature is as low as 0 °C, the water changes from liquid to solid (icing). When the temperature is higher than 0 °C, water changes from solid to liquid (dissolved). A large amount of cold energy is inhaled and stored during icing, while a large amount of thermal energy is absorbed during the dissolution process. The larger the amount (volume) of ice, the longer the dissolution process takes. This is one of the most typical examples of PCMs. PCMs are increasingly used in the electric heating industry. It represents a revolutionary transformation of traditional electric heating to energy-saving electric heating. Phase change thermoelectric heaters are high-performance representative products, which have the potential to save energy by 60~70% as compared with traditional electric heaters.

As can be seen from above examples, the PCMs can actually function as energy storage. This characteristic has a great significance in the field of energy saving and temperature control. Therefore, PCMs and their applications have become a wide research topic. The main differences between organic PCMs and inorganic PCMs are durability and fire resistance, etc., for which the latter is much better than the former.

PCMs have the ability to change their physical state over a range of temperatures. In the solid-liquid phase change, the change is observed at the phase change point. During the melting process, the PCM absorbs and stores a large amount of latent heat. When the PCM is cooled, the thermal energy is stored. The heat is released into the environment within a certain temperature range, and the reverse phase change process from liquid to solid is generated. In these two phase transitions, the energy stored or released is called the latent heat of phase change. When the physical state changes, the temperature of the material itself remains almost unchanged until the phase transition is completed. Although the temperature is constant, the latent heat absorbed or released is quite large. In these two phase transitions, the material absorbs heat from the environment and conversely, releases heat to the environment. When a large amount of phase change latent heat is transferred to the environment, a wide temperature platform is created. The appearance of PCMs embodies the extension of the constant temperature-time and can store or release sensible heat during thermal cycling. The principle is that the PCM stores energy during heat transfer, just like thermal resistance, which can extend the energy transfer time and can reduce the temperature gradient.

As the PCMs have the property of storing heat in the form of latent heat or releasing it to the environment during the phase change process, the PCMs can be introduced into the building envelope by appropriate design, and the outdoor temperature and the heat flow can be dealt with. In addition, the use of PCMs has the following advantages: first, the phase change process is generally an isothermal or nearly isothermal process, which is advantageous for maintaining temperature changes in a small range and making the human body feel comfortable; second, PCMs have high latent heat of phase change, a small amount of material can store a large amount of heat. Compared with sensible heat storage materials (such as concrete, brick, etc.), the structural requirements of the building can be greatly reduced, thereby, making the building able to adopt more flexible structures [2].

The classification of PCMs mainly includes inorganic PCMs, organic PCMs and composite PCMs. Among them, inorganic PCMs mainly include crystalline hydrated salts, molten salts, metals or alloys. Organic PCMs mainly include paraffin, acetic acid and other organic substances. CPCMs can effectively overcome the disadvantages of single inorganic or organic PCMs, improving the application effects of the material and expanding its application range. Therefore, the development of CPCMs has become a significant research topic in the field of thermal energy storage materials. However, CPCMs may also cause disadvantages such as a decrease in latent heat of phase change or a tendency to be denatured during long-term phase transition.

Overall, the use of PCMs is revolutionary in the field of construction. As mentioned earlier, the main result is energy saving, with energy savings reaching 60~99%. Taking the north heating in China as an example, where the PCMs are applied, the electricity consumption of a 100 m^2 house in a heating season is only about 10 kwh.

Shape stabilized composite phase change materials (SSCPCMs) have the advantages of both ordinary building materials and PCMs. These can absorb and release an appropriate amount of thermal energy and can be used simultaneously with other traditional building materials. In addition, no special knowledge or skills are required for installing or using SSCPCMs. These can be produced with standard production equipment, making these competitive in terms of economic efficiency.

The research on the application of CPCMs to building materials began in 1982, which was initiated by the US Department of Energy. In the 1990s, the technology of processing building materials such as gypsum boards, wallboards and concrete members by PCMs was developed. Subsequently, the research and application of PCMs in building materials such as concrete blocks and gypsum wallboards has been in ascendancy. In 1999, a new type of building material based on solid-liquid eutectic PCM was successfully developed. The PCM was cast in the wallboard or light concrete precast board to maintain the indoor temperature. In addition, there are a number of organizations in Europe and the United States that use PCM to produce outdoor communication wiring equipment and special transformers for power transformer equipment, which can be kept at suitable working temperatures in winter and summer. In addition, PCM containing asphalt or cement can prevent roads, bridges and airstrips from freezing in the winter.

The combination of PCM and building material matrix has the following general methods:

(1) Seal the PCM in a suitable container.
(2) The PCM is sealed and placed in the building material.
(3) The PCM is infiltrated into the porous building material substrate (such as gypsum wallboard, cement concrete test block, etc.) by immersion method.
(4) Mix the PCM directly with the building materials.
(5) Organic PCM is emulsified and added to the building material.

Several energy efficient domestic building enterprises have successfully emulsified paraffin of different types, and subsequently mixed with phase change special rubber powder, water, polyphenyl granule light aggregate in a certain proportion, which can be used for a phase change heat storage slurry of the inner and outer layers of building wall for both heat storage and heat preservation. At the same time, there are also products such as phase change mortar and phase change putty, among others.

Modern buildings are ever-increasing in height, requiring the envelopes used to be lightweight materials. However, ordinary lightweight materials have a smaller heat capacity, resulting in greater indoor temperature fluctuations. This not only causes the indoor thermal environment to be uncomfortable, but also increases the air-conditioning load, resulting in an increase in building energy consumption. The latent heat of the phase change material reaches 170 J/g or higher, while the ordinary building materials would require 190 times the quality of the PCM when storing the same heat at a temperature change of 1 °C. In this respect, the shape stabilized phase change materials (SSCPCMs) have the heat capacity unmatched by ordinary building materials and are very favorable for the indoor temperature stability and the smooth operation of the HVAC system. The disadvantage of this composite product is that it is not strong enough, thus, limiting its capacity to hold much weight. CPCMs consist of support materials and PCMs, which are described in the below sections.

3.2.2 Support Materials

In terms of support materials, porous, lightweight, chemically stable and structurally strong materials such as expanded perlite, expanded vermiculite, gypsum, concrete and porous aluminum are commonly used.

Expanded Perlite (EP)

Expanded perlite (Figure 3.1) is a kind of natural acidic vitreous lava, which is a non-metallic mineral including perlite, rosin and obsidian, all of which differ in crystal water content. As it expands rapidly 4~30 times under high temperature conditions of 1000~1300 °C, it is collectively referred to as expanded perlite. All of these minerals are exploited by open pit mining. There is no need for beneficiation, just crushing and sieving is appropriate. EP can be used as a filter, catalyst, molecular sieve and carrier for rubber, fertilizer, pesticide, etc. It is widely used in construction, metallurgy, petroleum, machinery, light industry, hydropower, foundry, medicine, food, agriculture, forestry and horticulture.

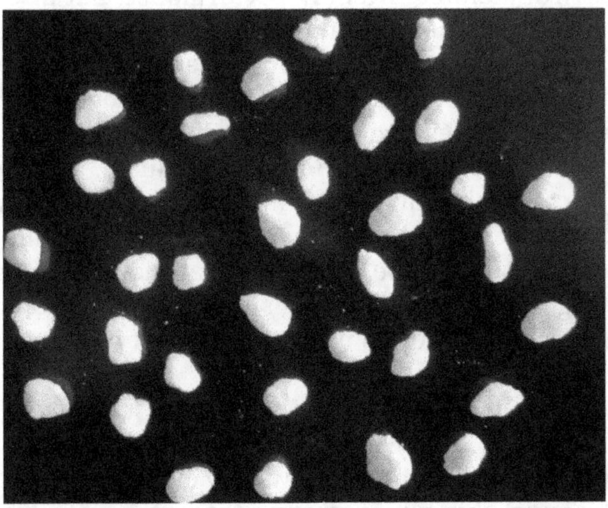

Figure 3.1 Expanded perlite.

EP is a white granular material with a honeycomb structure, inside which the perlite ore is preheated and expanded by instantaneous high temperature roasting. The principle is as follows: the perlite ore is crushed to form a certain size of ore, preheated and calcined, followed by rapid heating (above 1000 °C). The water in the ore is vaporized and expands inside the softened vitreous ore to form a porous structure and enlarges in volume by 10~30 times of non-metallic mineral products. Perlite is divided into three forms according to its expansion process technology and application: open cell type, closed cell type, and balloon type.

Cons: EP has high water absorption rate and poor water resistance, which leads to large volume shrinkage deformation of the thermal insulation mortar during mixing, low thermal insulation performance, easy cracking and low adhesion strength to the base layer. EP thermal insulation mortar is prone to delamination, segregation and bleeding. The on-site construction performance is poor, and the technical performance of the thermal insulation mortar after hardening is affected.

Pros: EP has excellent thermal insulation performance and good stability. It is well accepted by the and has a wide range of applications with universal practicability, especially in terms of refractory insulation and energy-saving. It is widely used not only because of its exceptional environmentally friendly performance, but also due to its economic and social potential value, for example, construction expediency, easy maintenance, etc. Its impact performance is superior to any other insulation protection. Under normal conditions of use as well as in case of dryness, severe cold, high temperature, humidity, galvanic corrosion or the growth of insects, fungi, algae, etc., along with damage caused by sawtooth, impact of objects, etc., no damage would is caused to EP and the service life of the building is greatly extended. The lifespan of the insulation material is the same as that of the building. Due to its multi-porosity, stable chemical properties and long service life, it is often used as a support material for CPCMs.

Expanded Vermiculite (EV)

Vermiculites (Figure 3.2) have been widely used in construction, metallurgy, petroleum, ship building, environmental protection, thermal insulation, fireproofing, energy conservation and other fields. Vermiculite has unique structural and surface properties as well as non-toxic, sterile and chemically inert nature. It can be used as a carrier, adsorbent, anchor and feed additive.

After the raw vermiculite is calcined at a high temperature, the volume can rapidly expand several tens of times, and the volume-expanded vermiculite is called expanded vermiculite. EV series products are golden yellow vermiculite and silver-white vermiculite;

varieties include vermiculite tablets, vermiculite powder, horticultural meteorites, mixed expansion vermiculite, etc.

Figure 3.2 Expanded vermiculite.

Vermiculite is generally divided into vermiculite ore and expanded vermiculite. The volume of vermiculite ore can rapidly expand by high temperature roasting up to 8~20 times, and the specific gravity after expansion is 130~180 kg/m^3, which has strong thermal insulation performance.

Gypsum

Gypsum is a monoclinic mineral and a hydrate with calcium sulfate ($CaSO_4$) as the main chemical component. Gypsum is a versatile industrial and building material. It can be used in cement retarders, gypsum construction products, model making, medical food additives, sulfuric acid production, paper filler, paint filler, etc.

The microporous structure and heat dehydration of gypsum and its products make it excellent in sound and heat insulation as well as fire resistance. In general, gypsum refers to two kinds of minerals, gypsum and anhydrite. The raw gypsum is calcium sulfate dihydrate ($CaSO_4 \cdot 2H_2O$), also known as dihydrate gypsum or soft gypsum. The crystal is plate-like, usually in the form of dense block or fiber, with (a) white, gray, red or brown color, (b) glass or silk luster, (c) Mohs hardness 2 and (d) density 2.3 g/cm^3. Anhydrite is anhydrous calcium sulfate ($CaSO_4$), with theoretical composition CaO 41.2% and SO_3 58.8%. The orthorhombic crystal is plate-like, with (a) usually dense blocky or granular morphology, (b) white or off-white color, (c) glass luster, (d) Mohs hardness of 3 to 3.5 and (d) density 2.8 to

3.0 g/cm[3]. The two types of gypsum are often produced, which can be transformed into each other under certain geological conditions.

After calcination and grinding natural dihydrate gypsum ($CaSO_4 \cdot 2H_2O$), β-type hemihydrate gypsum ($2CaSO_4 \cdot H_2O$), that is, building gypsum, also known as plaster, is formed. If the calcination temperature is 190 °C, model gypsum can be obtained, and its fineness and whiteness are higher than construction gypsum. If the gypsum is calcined at 400~500 °C or even higher, the floor gypsum will be solidified and hardened, but the strength, wear resistance and water resistance after hardening are better than ordinary building gypsum.

Gypsum is usually white, colorless and transparent crystal, which is also called transparent gypsum. Occasionally, due to the presence of impurities, it would usually present gray, light yellow, light brown and other colors. It has often white stripes with transparent glassy luster or fibrous aggregate silky luster.

Concrete

Concrete refers collectively to engineering composite materials in which aggregates are cemented into a whole by cementitious materials. Generally speaking, the term concrete refers to cement as cementing material, sand and stone as aggregate. Cement concrete, also called ordinary concrete, is blended with water (which may contain admixture) and stirred, and is widely used in civil engineering.

The inorganic cementitious materials such as cement, lime and gypsum are mixed with water to make the concrete mixture plastic. Subsequently, various chemically and physicochemical processes produce strength by coagulation hardening. In general, drinking water can meet the requirements of concrete mixing water. Excessive amounts of acids, alkalis, salts and organics in the water can have a detrimental effect on concrete. Aggregates not only have a filling effect, but also have an important influence on the bulk density, strength and deformation properties of concrete.

In order to improve certain properties of concrete, an admixture may be added. Due to the obvious technical and economic effects of admixtures, it is increasingly becoming an indispensable component of concrete. In order to improve the workability of the concrete mixture or the performance of the concrete after hardening, admixture can be incorporated during the concrete mixing, thus, leading to saving of cement. It is divided into two categories, active and inactive.

The nature and quantity of the admixture affect the strength, deformation, hydration heat, impermeability and color of concrete.

Porous Aluminum

Porous aluminum [3] is a metal-based composite functional material that is integrated with the structure and function of aluminum or aluminum alloy matrix phase and gas phase, which is uniformly distributed in the solid phase. Porous aluminum has excellent properties such as lightweight, low density, large specific surface area, high heat dissipation capacity, sound absorption and insulation, strong filtration capacity, high temperature resistance, no pollution, shock absorption and damping, etc. [4-5].

According to different pore structures, porous aluminum can be categorized into open cell and closed cell types. Open cell aluminum can be used as filter material for high temperature liquid filtration and dust removal [6], as well as support for some soft materials, etc. Aluminum can be used as a sandwich interlayer for weight reduction and cushion of mechanical equipment [7]. In recent years, there have been many related research studies on this aspect.

3.2.3 PCMs

There are many ways to classify PCMs, based on phase transition temperature, chemical composition and phase change process. According to the phase transition temperature, PCMs can be divided into high temperature, medium temperature and low temperature PCMs [12]; on the basis of chemical composition, PCMs can be sorted as organic, inorganic and composite PCMs [13]; whereas on the basis of the phase change, PCMs can be categorized as solid-liquid, solid-solid, solid-gas and liquid-gas phase change materials [14]. The classification of PCMs is also depicted in Figure 3.3.

For the development of CPCMs for building envelopes, it is important to select suitable PCMs, which should have the following characteristics [15-18]:

(1) The latent heat of fusion should be high, so that the material can store or release more heat during the phase change;
(2) The phase change process should have good reversibility, small expansion and contraction, along with low extent of sub-cooling or overheating;

(3) The material should be of a suitable phase transition temperature to meet the specific temperature to be controlled;
(4) The thermal conductivity, density and specific heat capacity should be high;
(5) The PCM should be non-toxic, non-corrosive, low in cost and convenient to manufacture.

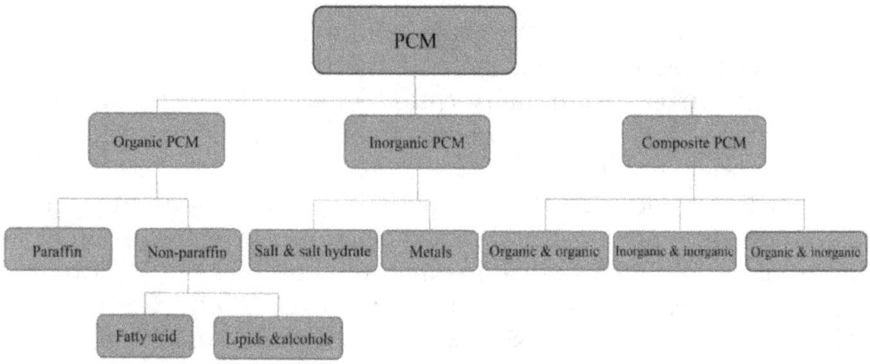

Figure 3.3 Classification of PCMs.

A brief introduction of a few important PCMs is presented in the following sections.

Organic PCMs

Organic solid-liquid PCMs mainly include aliphatic hydrocarbons, fatty acids, alcohols and polyenols. The advantages include: phase separation and supercooling do not readily occur, low extent of corrosion and large latent heat of phase change [19]. The main disadvantage is the relative ease of leakage. At present, most of the applications are mainly based on aliphatic hydrocarbons and polyhydric alcohol compounds. The organic solid-solid PCMs store or release energy through the conversion of the crystal form of the material, and have the advantages of small volume change, no leakage, no corrosion and long service life in the phase change process. Three main types of solid-solid PCMs have been developed with economic potential: polyols, polymers and layered perovskites. Organic PCMs also include paraffin waxes and non-paraffins, among which non-paraffins are also classified into fatty acids as well as lipids and alcohols.

There are numerous kinds of organic PCMs. However, for building energy conservation, the phase transition temperature is required to be within a suitable temperature range, generally 20~40 °C. When the phase transition temperature of an organic substance is beyond this range, several organic substances may be combined to form a multi-component organic PCM, which is adjusted to have a suitable phase change temperature and latent heat. Compared with solid-liquid PCMs, solid-solid PCMs have many advantages such as no liquid in the phase change process, small volume change, no need for container storage and easy processing. Therefore, widespread fundamental and applied research has been carried out on this aspect in recent years [20-22].

Paraffin Organic PCMs

Paraffin organic PCMs are mainly divided into edible wax, fully refined paraffin, semi-refined paraffin and crude paraffin. The short-chain alkane such as ethane (C_6H_{14}) has a low melting point of -95.4 °C. As the number of the carbon atoms in the chain increases, it begins to grow faster, and subsequently tends to level off (melting point of $C_{30}H_{62}$ is 65.4 °C and $C_{40}H_{82}$ is 81.5 °C). As the chain grows in length, the latent heat of fusion of alkane also increases [23].

Paraffin wax, also known as crystalline wax, is a mixture of hydrocarbons with a carbon number of about 18~30. Its mainly consists of a linear alkane (about 80~95%), along with a small amount of alkane with individual branches and long side chains. Paraffin wax crystals are flaky or needle-like, which are obtained by solvent refining, solvent dewaxing or wax freezing crystallization, deoiling and refining.

Each type of wax is divided into different varieties according to the melting point, generally using a 2 °C period, such as No. 52, No. 54, No. 56, etc. Crude paraffin has high oil content and is mainly used for manufacturing matches, fiberboards, tarpaulins, etc. Fully refined paraffin and semi-refined paraffin are widely used as food and oral medicines, components and packaging materials for certain commodities (such as wax paper, crayons, candles, carbon paper, etc.), coating materials for baking containers, fruit preservation agents, insulation for electrical components, agents to improve rubber aging resistance and flexibility, etc. It can also be used to oxidize synthetic of fatty acids.

The hydrocarbons in paraffin wax are extracted from certain distillates of petroleum, shale oil or other bituminous mineral oil. The solid alkanes, forming the main component, are odorless, tasteless and white or light yellow translucent solid. Paraffin is amorphous, but can have distinct crystal structure. Paraffin wax is usually a white, and odorless waxy solid, with melting point in the range of 47~64 °C, density of about 0.9 g/cm^3 and solubility in gasoline, carbon disulfide, xylene, ether, benzene, chloroform, carbon tetrachloride, naphtha, etc. Paraffin wax is also a good heat storage material with a specific heat capacity of 2.14~2.9 J/(g·K) and a heat of fusion of 200~220 J/g. The main performance indicators of paraffin are melting point, oil content and stability [24].

As paraffin is a mixture of hydrocarbons, thus, it does not have a strict melting point like a pure compound. The melting point of paraffin wax refers to the temperature at which the paraffin sample is cooled and melted under the specified conditions, and the stagnation period occurs for the first time on the cooling curve. All kinds of wax products require good temperature resistance, that is, they do not melt or soften below a specific temperature. Commercial paraffin waxes are required to have a range of different melting points depending on the conditions, region, season of use and environment in which they are used. The main factor affecting the melting point of paraffin is the weight of the selected raw material, and the melting point of the paraffin wax removed from the heavier fraction is higher. In addition to that, the oil content also has a great influence on the melting point of the paraffin wax. As a PCM for the construction industry, the phase change temperature of paraffin wax is tuned between 25~30 °C.

Table 3.1 also presents physical properties of some common paraffins.

Table 3.1 Physical properties of some paraffins

Paraffin	Freezing point/range (°C)	Heat of fusion (kJ/kg)
25#	24-25	106.2 [26]
6106	42-44	189 [25]
P116	45-48	210 [25]
5838	48-50	189 [25]
6035	58-60	189 [25]
Paraffin wax	64	173.6 [27]
Paraffin C21-C50	66-68	189 [30]

Non-paraffin Organic PCMs

Non-paraffin PCMs include fatty acids and alcohols.

Fatty acids: Fatty acid refers to a long aliphatic carbon chain containing a carboxyl group The long-chain carboxylic acids can be saturated (without double bonds) or unsaturated (with double bonds). Most of these are straight chain molecules, while some of them also have branched chains. The lower fatty acid is a colorless liquid with pungent odor. The higher fatty acid is a waxy solid with no obvious odor.

Promising fatty acid PCMs such as stearic acid, palmitic acid, myristic acid and lauric acid have broad prospects for phase change energy storage, but the high cost limits their practical applications. The combination of composite sheets and fatty acids can effectively reduce the cost and improve the performance of its architectural applications.

Lipids: Lipids, also known as adipoids, are a general term for fats. These are non-polar organic solvents that are insoluble in water and soluble in fatty solvents (alcohols, ethers, chloroforms and benzene). They are important organic compounds that can be used by human bodies. Lipids are usually classified according to the main components: simple lipids, complex lipids, derived lipids and non-saponified lipids. Lipids include a wide variety of molecules characterized by a non-polar covalent bond consisting primarily of both carbon and hydrogen. Strictly speaking, lipids are not macromolecules because their relative molecular mass is not as high as sugars, proteins and nucleic acids, and they are not polymers.

Alcohol: An alcohol is a compound having a hydroxyl group bonded to a hydrocarbon group or a carbon on a side chain of a benzene ring. Important alcohols are methanol, ethanol, benzyl alcohol, ethylene glycol, etc. A class of organic compounds is formed by the substitution of one or more hydrogens in a hydrocarbon molecule by a hydroxyl group. A compound formed by substituting hydrogen on a ring of an aromatic hydrocarbon with a hydroxyl group is not an alcohol but a phenol.

Ethanol and other alcohols are present in the fermentation broth. Various sterols and aromatic alcohols are found in the plant essential oils, which exist in a free state or in the form of esters and acetals. Still many alcohols are present in the form of esters in animal and vegetable oils, fats and waxes. According to the number of hydroxyl groups contained, the family of alcohols can be divided into monohydric, dihydric, trihydric or polyhydric alcohols. The alcohols may also be sorted into primary, secondary and tertiary alcohols on the basis of the number of hydrogens on the carbon atom to which the hydroxyl group is attached.

Table 3.2 also presents the melting points and latent heats of some common non-paraffins.

Table 3.2 The melting points and latent heats of some non-paraffins

Material	Melting point (°C)	Latent heat (kJ/kg)
Diethylene glycol	-10	247 [28]
Triethylene glycol	-7	247 [28]
Formic acid	7.8	247 [29]
Glycerin	17.9	198.7 [25]
D-Lattic acid	26	184 [25]
Methyl palmitate	29	205 [25]
Docasyl bromide	40	201 [25]

Inorganic PCMs

Inorganic solid-liquid PCMs mainly include crystalline hydrated salts, molten salts and metal alloys. Most of the crystalline hydrated salts are medium-low temperature PCMs, mainly including halides, sulfates, phosphates, nitrates, acetates, carbonate hydrates of alkali metals and alkaline earth metals. Such PCMs are inexpensive with large volumetric heat storage density and high phase change heat value. They have a thermal conductivity larger than that of organic PCMs and are generally neutral. However, in use, they are prone to unfavorable factors such as sub-cooling and phase separation. Their thermal storage performance is unstable [30-31].

The molten salts (mainly fluoride, chloride, oxide, carbonate, nitrate, etc.) of some alkali metals have generally high phase change temperature, but the thermal conductivity is low. At the same time, these phase change materials are highly corrosive at high temperatures [12-14].

Metals or alloys have large latent heat, high thermal conductivity, large heat storage density, small volume change during phase change and no sub-cooling, however, they lead to corrosion at high temperature [15-17].

Inorganic solid-solid PCMs mainly include layered perovskites, KHF_2, NH_4SC, etc. Their phase change latent heat values are high, along with small degree of sub-cooling and good thermal stability.

Mixed Phase Change Materials

Mixed phase change materials include organic/organic CPCMs, inorganic/inorganic CPCMs and organic/inorganic CPCMs.

For instance, the organic/inorganic SSCPCMs are composed of an organic PCM and an inorganic carrier and the heat is stored or released by the exothermic and endothermic heat of the organic PCM in the solid-liquid phase change process. Compared with common PCMs, these do not require packaging equipment, which reduces packaging cost and packaging difficulty, avoids material leakage defects, reduces heat transfer resistance of the container and facilitates heat exchange between the PCM and the heat transfer fluid.

In order to improve the supercooling of PCMs while maintaining high latent heat and thermal stability, Wu *et al.* [32] prepared different ratios of stearic acid/$Mg(NO_3)_2{\cdot}6H_2O$ and stearic acid/$Na_2HPO_4{\cdot}12H_2O$ mixed materials. Using alternating high/low temperature to test long-term cycle performance and taking advantage of a temperature recorder to measure the step cooling curve to obtain phase transition temperature and sub-cooling, the phase change latent heat before and after the thermal cycle was tested and compared. Results showed the degree of sub-cooling of the stearic acid/$Na_2HPO_4{\cdot}12H_2O$ mixed material reduced to about 3 °C. After 300 cycles of melting/solidification, the degree of sub-cooling was constant and the latent heat decay rate was within 20%. The authors concluded that the two kinds of CPCMs with similar structure can improve the degree of sub-cooling of inorganic hydrated salts. The compatibility of stearic acid with $Mg(NO_3)_2{\cdot}6H_2O$ was not optimal, thus, the decay of latent heat of phase change was intensified and cycle stability was deteriorated. The compatibility of stearic acid with $Na_2HPO_4{\cdot}12H_2O$ was good, which along with the stable performance made it an excellent CPCM [32].

Table 3.3 also lists the properties of some inorganic PCMs [33-46].

3.3 Enhancement of Thermal Conductivity of PCMs

In summary, the most suitable materials among the family of PCMs for architectural applications are organic PCMs. However, the organic

Table 3.3 Properties of potential inorganic PCMs

Compound	Melting temperature (°C)	Heat of fusion (kJ/kg)	Thermal conductivity (W/m·K)	Density (kg/m³)
H_2O	0 [33,35]	333 [33]	0.612 (liquid, 20 °C) [33]	998 (liquid, 20 °C) [33]
		334 [35]	0.61 (30°C) [35]	996 (30 °C) [35]
				917 (solid, 0 °C) [33]
$LiClO_3·3H_2O$	8.1 [42,43]	253 [42]	n.a.	1720 [42]
$ZnCl_2·3H_2O$	10 [42]	n.a.	n.a.	n.a.
$K_2HPO_4·6H_2O$	13 [38]	109 [25]	n.a.	n.a.
$Na_2CrO_4·10H_2O$	18 [42]	n.a.	n.a.	n.a.
$Mn(NO_3)_2·6H_2O$	25.8 [38]	125.9 [40]	n.a.	1738 (liquid, 20 °C) [40]
				1728 (liquid, 40 °C) [40]
				1795 (solid, 5 °C) [40]
$FeBr_3·6H_2O$	27.0	105 [25]	n.a.	n.a.
$CaCl_2·6H_2O$	29 [34,42]	190.8 [34,42]	0.540 (liquid, 38.7 °C) [34,42]	1562 (liquid, 32 °C) [34,42]

PCMs have the disadvantage of poor thermal conductivity, which affects their application in building maintenance to some extent. For these PCMs, the phase transition temperature is lower and the phase transition enthalpy is larger, which is suitable for application for building envelope structure. As low thermal conductivity directly affects their thermal storage performance, any research efforts have been made for enhancing the thermal conductivity of PCMs.

In recent years, research studies on CPCMs have reported new developments in this direction. CPCMs can not only overcome the corrosive properties of PCMs, low phase transition temperature, small phase change enthalpy, large volume change during phase change process, long reaction time to environmental changes, but can also effectively interact with supporting materials. In order to enhance the thermal conductivity, a material with a large thermal conductivity, such as aluminum powder, copper powder and expanded graphite, is generally incorporated with the CPCM.

3.3.1 Aluminum Powder

The content of aluminum in the earth's crust is second only to oxygen and silicon. Thus, it is the most abundant metal element in the earth's crust. Aluminum powder, commonly known as "silver powder" (or silver metallic pigment), is generated by adding a small amount of lubricant in pure aluminum foil, crushing into scaly powder by smashing and subsequently polishing. Aluminum powder is light in weight, high in floating force, strong in hiding power and good in reflection performance against light and heat. After treatment, it can also be a non-floating material. Aluminum powder can be used to identify fingerprints and can also be used in fireworks. In addition, it is a widely applied metallic pigment.

Aluminum powder is easy to prepare and is inexpensive. Due to its high thermal conductivity, it is suitable as an additive for CPCMs to enhance the thermal conductivity. Thus, CPCMs can effectively perform the heat absorption and release process, thus, enhancing regulation and mitigating the effects of outdoor temperature on the indoor environment.

Mettawee *et al.* [47] incorporated aluminum powder with particle size of 80 μm in a CPCM (paraffin-based) at mass ratios of 0.1, 0.3, 0.4 and 0.5, respectively. The experimental results showed that after aluminum powder blending, the endothermic process time of the composite paraffin PCM was shortened by nearly 60% compared with the paraffin CPCM without aluminum powder.

Aiming at the shortcomings of paraffin wax as a solid-liquid PCM such as low thermal conductivity and poor heat transfer performance, Yang *et al.* [48] prepared a nano-aluminum powder/paraffin CPCM with good dispersibility by two-step method. Investigation of the thermal properties revealed that the addition of nano-aluminum powder effectively improved the thermal conductivity of paraffin CPCMs, but had a little effect on the latent heat of phase change and phase transition temperature [48].

Similarly, other recent studies have also revealed the usefulness of aluminum powder in enhancing the thermal conductivity of CPCMs, among other properties.

3.3.2 Copper Powder

Pure copper is a soft metal with a reddish-orange metallic luster. The element is purple-red, when the surface is freshly cut. It has excellent ductility, high thermal conductivity and brilliant electrical conductivity. Therefore, it is the most commonly used material in cables as well as electrical and electronic components. It can also be used as a building material to form a wide variety of alloys. Copper alloys have excellent mechanical properties and low electrical resistivity, the most important of which are bronze and brass. In addition, copper is also a durable metal that can be recycled multiple times without compromising its mechanical properties.

In view of the phase separation, super-cooling problem and low thermal conductivity of hydrated salt PCMs, Li *et al.* [49] used copper powder and sodium trihydrate to generate the CPCM. Experiment studies showed that the CPCM was superior than pure PCM. The thermal conductivity of the salt PCM was improved by eleven times, with good thermal stability, low sub-cooling and high thermal conductivity [49].

Zhang [50] reported ultrasonic dispersion of nano-copper powder and dispersant Htenol BC-10 to obtain nano-copper-paraffin CPCM. The thermal conductivity and energy efficiency were greatly improved. The addition of nano-copper powder had a little influence on the phase transition temperature. On the other hand, thermal stability and thermal conductivity of the CPCM were effectively improved on the addition of the nano-copper powder.

3.3.3 Expanded Graphite (EG)

As a new class of functional carbon material, expanded graphite (EG) is a loose porous substance obtained by intercalating, washing, drying and high-temperature expansion of natural graphite scales. In addition to the excellent properties of natural graphite itself, such as resistance to cold, heat, corrosion and self-lubrication, EG also has compression resilience, adsorption, ecological environment coordination, biocompatibility and radiation resistance, which are not found in natural graphite. As early as the early 1860s, Brodie heated natural graphite with chemicals such as sulfuric acid and nitric acid and observed expanded graphite. However, its application began only recently. Since then, many research on EG and has led to many significant scientific breakthroughs. At high temperature, EG can expand in volume by 150~300 times, from sheet to worm-like, such that the structure is loose, porous and curved, with the enhancement in surface area and surface energy. This increases its softness, resilience and plasticity.

As mentioned earlier, the PCM has low thermal conductivity and poor heat transfer performance, which affects its energy storage and energy release efficiency. In addition, the porosity of the porous medium in the PCM is small, resulting in low energy storage. These shortcomings limit the application and development of the material. The rich porous structure EG and high thermal conductivity can make up for these defects.

Zhang *et al.* [51] directly adsorbed paraffin on EG, thereby preparing paraffin/EG CPCM with different mass ratios and studied the storage and exothermic properties of paraffin/EG CPCM. For a paraffin content of 80%, the storage and exothermic time of the CPCM were 69.7% and 80.2% shorter than that of the paraffin respectively, and the CPCM had good cycle performance heat stability.

Zhao *et al.* [52] used the adsorption properties of porous EG structure to prepare polyethylene glycol/EG CPCM. The thermal properties of the material were studied by differential scanning calorimetry (DSC). The results showed that the phase transition temperature of the composite did not change with the variation of polyethylene glycol content and the phase change latent heat of the material increased with the augmentation of polyethylene glycol fraction. On the other hand, the thermal conductivity of the composite decreased with growing polyethylene glycol content. The porous structure of EG had good adsorption properties for polyethylene glycol. No seepage of liquid polyethylene glycol was observed during the solid-liquid phase change.

3.3.4 Carbon Nanotubes

Carbon nanotubes are light in weight, with perfectly connected hexagonal structure and have many superior mechanical, electrical and chemical properties. In recent years, with significant research focus on carbon nanotubes and nanomaterials, broad application prospects of these materials have been revealed.

Carbon nanotubes, also known as bucky tubes, are one-dimensional quantum materials with a special structure (radial size in the order of nanometers, axial dimension in the order of microns and substantially sealed at both ends). The carbon nanotubes are mainly composed of a plurality of coaxial tubes with carbon atoms arranged in a hexagonal shape. According to the different orientation of the carbon hexagon in the axial direction, nanotubes can be divided into three types: zigzag, armchair and spiral.

The photoelectron spectroscopy results on carbon nanotubes show that regardless of single-walled or multi-walled carbon nanotubes, there are certain functional groups on the surface of the material and the carbon nanotubes have different surface structures due to the different post-treatment processes and preparation methods. In general, single-walled carbon nanotubes are highly chemically inert with pure surface and the surface of multi-walled carbon nanotubes is much more active, combined with a large number of surface groups, such as carboxyl groups.

Carbon nanotubes have good heat transfer performance, with a very large aspect ratio. Their thermal conductivity along the axial direction is high and the same property of the material in the radial direction is relatively low. Nanotubes can be synthesized to prepare highly anisotropic heat conductive materials.

Xie *et al.* [53] used a ball milling method to fabricate multi-walled carbon nanotube-paraffin CPCM. The thermal conductivity of the compound was increased by 35% and 40% respectively, in the solid and liquid phases, with 2% mass fraction of carbon nanotubes [53].

3.3.5 Graphene Wafers

Graphene is a two-dimensional carbon nanomaterial with sp^2 hybridized carbon atoms forming a hexagonal honeycomb lattice. Graphene has excellent optical, electrical and mechanical properties, thus, has high potential of application in materials science, micro-nano processing, energy, biomedicine and drug delivery.

The pure defect-free single-layer graphene has a thermal conductivity of up to 5300 W/m.K, which is the highest thermal conductivity value among carbon materials, higher than single-walled carbon nanotubes (3500 W/m.K) as well as multi-walled carbon nanotubes (3000 W/m.K). The thermal conductivity can reach 600 W/m.K in case of its use as a carrier [54].

Wang *et al.* [55] used transient hot-wire thermal conductivity measurement and DSC to characterize the thermal conductivity and phase change characteristics of different mass fractions of paraffin-based nano-CPCMs prepared by using multi-walled carbon nanotubes as filler. The experimental results showed that although the phase transition temperature of the CPCM was almost constant, the phase transition enthalpy decreased with an increase in the carbon nanotubes content. As the mass fraction of carbon nanotubes was increased to 5%, the phase change enthalpy was about 15% lower than that of pure paraffin. At 30 °C and 50 °C, the thermal conductivity had a uprush due to the impact of the solid-solid and solid-liquid phase changes. In addition, the thermal conductivity showed a linear growth trend with the mass fraction of nanotubes. At a mass fraction of 5%, the maximum relative improvement rate was close to 40%, showing a good strengthening effect [55].

3.4 Preparation of CPCMs

With the continuous development of science and technology in general and PCM knowhow in specific, the methods of generating the CPCMs have also evolved. This section briefly introduces several preparation methods for CPCMs, including sol-gel method, infiltration method, vacuum infiltration method, drum rotating infiltration method, etc.

3.4.1 Sol-gel Method

Sol-gel technology refers to a method in which a metal organic or inorganic compound is gelated and heat-treated to form an oxide or other solid compound. The process involves using a liquid chemical reagent or a sol as a raw material, mixing uniformly in a liquid phase and performing a reaction to form a stable and precipitate-free sol system, which would turn into a gel after being left for a period of time. Subsequently subjecting to dehydration treatment in a sol or gel forms the product. The process is characterized by mild reaction conditions and uniform dispersion of two phases. Various polymer-based nanocomposites with different properties have been prepared using this mehtod by changing the reaction components [56].

The most common sol-gel method is the encapsulation of PCMs by ludox. Ethyl orthosilicate is used as precursor, ethanol is used as solvent, and hydrolysis and crosslinking occur simultaneously under the action of mechanical stirring and hydrochloric acid. The silica sol is obtained, which after the aging process, forms a gel with a three-dimensional network structure having a "cage structure". By adding a PCM to the reaction system, the PCM can be effectively bound in the "cage structure" and is not easily leaked, even in liquid state. The reaction conditions are mild and the reaction can proceed under normal temperature and pressure, indicating a good development potential [57].

Ren *et al.* [58] used sol-gel method to prepare CPCMs, prepared by using silica powder, floating bead and activated carbon as adsorbent materials and PCMs. The results of scanning electron microscopy (SEM) and DSC analyses indicated that the CPCMs generated using activated carbon and PCMs had higher phase transition enthalpy, more suitable phase transition temperature and better thermal stability.

Zhang [59] used tetraethyl orthosilicate, water, absolute ethanol and stearic acid as raw materials for the sol-gel synthesis. X-ray diffraction, DSC, SEM and infra-red (IR) analyses were utilized to characterize the prepared CPCM. The experimental results showed that the sol-gel method was simple to carry out and the prepared sample was relatively uniform with high purity. The prepared porous CPCM had lower phase transition temperature, however, the heat storage capacity was improved, as compared with the original PCM. The phase change latent heat value of the developed CPCM with stearic acid content of 35% reached 64.23 J/g, and the phase transition temperature was 51.81 °C, confirming the stability of the material [59].

3.4.2 Infiltration Method

Melt infiltration usually results in two kinds of CPCM materials with different melting points. The high melting point material is used as a support to prepare a porous matrix with a network structure. The low melting point is the PCM melted into the porous matrix. In order to improve the thermal conductivity, a heat conductive material such as graphite can be added, and this method is suitable for preparing a low-temperature SSCPCM for industrial and construction uses [60].

The melt infiltration method generally adopts a melt pressure-less infiltration process. Pressure-less infiltration has the following requirements for the PCM melt and porous matrix: (1) PCM melt should cope with porous matrix infiltration; (2) matrix should have interconnected infiltration channels; (3) system component properties should match and (4) Infiltration conditions should not be harsh [61].

Huang [62] analyzed the static and dynamic conditions of spontaneous melting impregnation process, from the perspective of thermodynamics. The author established dynamic model of the infiltration of molten salts Na_2SO_4/SiO_2, by theoretically determining various influencing factors of the impregnation process and the relationship between them. Thus, the author was successful in forming a relatively complete inorganic salt/ceramic CPCM spontaneous melting infiltration theory [62].

Infiltration method is based on porous solid adsorbent to let one or several of the liquid components adsorbed on the surface, with a suitable solvent, as well as heating or blowing to let the component desorption to be carried out, thus, achieving the purpose of separation and enrichment of the components.

Taking advantage of the characteristics of small pores inside the porous medium, dispersion of the PCM by means of capillary effect improves the reliability of the PCM in porous media storage and avoids leakage in the process of solid-liquid phase change. Also, it allows to make use of the characteristics of high thermal conductivity of porous media to improve the thermal conductivity of the PCM. Thus, for selecting a porous medium, its structural characteristics (pore diameter distribution, pore shape and hole connectivity) and compatibility with PCMs must be taken into account for effective performance.

3.4.3 Vacuum Infiltration Method

The vacuum infiltration method has been developed for the preparation of composites of PCMs and porous media. It is generally difficult for the porous materials to absorb large amounts of the liquid, owing to the large amount of compressed air stranded inside the material's internal pores, which hinders the liquid to penetrate inside. In this respect, a large number of organic PCMs can be loaded into porous media by vacuum infiltration. After the organic PCMs loaded porous media particles are dried and cleaned on the surface, a simple dip coating method can be adopted to add a low permeability coating on the surface of the particles, in order to attain leakage-free CPCMs [63].

Lafdi *et al.* [64] combined carbon foam with PCM to prepare CPCM using vacuum infiltration method. The results showed that the combination led to CPCM with higher thermal conductivity than pure PCM. For aerospace applications, the phase change enthalpy was observed to increase more than 8 times. In civil applications, using foamed carbon with a porosity of 97% combined with PCMs, the average phase change enthalpy of the developed CPCM could be increased by 5 times as compared to pure PCM.

3.4.4 Drum Rotating Infiltration Method

In order to enhance the incorporation of the PCM with the support material and to improve the output of the CPCM, a new infiltration method, named as drum rotating infiltration method, has been developed. This method requires a rotating drum roller, an electric heater, a vacuum pump, a motor and other necessary equipment. Its principle is to let the PCM infiltrate after it is heated to melting point and the air inside the micro-structure of the pores is evacuated by the vacuum pump, while the drum roller should be rotating during the experiment.

Kong *et al.* [65] prepared SSCPCM particles by combining paraffin and expanded perlite using drum rotating infiltration method and obtained the optimal ratio between paraffin and expanded perlite. The thermal properties, morphology and chemical stability of the SSCPCM particles were analyzed through DSC, SEM and FT-IR tests. The experimental results showed that the SSCPCM could effectively enhance the inertia of the envelope structure of the experimental room, thus, improving the indoor thermal comfort.

3.5 Properties

3.5.1 Thermal Properties

In DSC, the relationship between the power difference (in the form of heat, for example) and temperature of the sample and reference material is measured at the controlled temperature. Also, DSC can be used to measure a variety of thermodynamic and kinetic parameters, such as specific heat, heat of reaction, heat transformation, phase diagram, reaction rate, crystallization rate, polymer crystallinity and purity of samples. The method can adapt a wide temperature range and high resolution, with low sample consumption. It is suitable for inorganic substances, organic compounds and drug analysis. For CPCMs, DSC is mainly used for the thermal property analysis, such as phase change enthalpy of the PCM, latent heat of the material, phase change temperature, sub-cooling and other thermal parameters of PCM.

Sari *et al.* [66] carried out DSC analysis to investigate the thermal properties of paraffin/EG CPCM. The experiments showed that the latent heat capacity of the CPCM was correlated to the ratio of paraffin in the CPCM. The more paraffin ratio (mass ratio 2% to 10%) was present in the CPCM, the larger the latent heat of the CPCM was observed. Besides, the higher the thermal conductivity of the CPCM, the shorter was the phase change process. The total values of latent heat of the CPCMs were large and almost equivalent to values calculated by multiplying the latent heat value of pure paraffin (194.6 kJ/kg) with its mass fractions (98, 96, 93, and 90 wt%). In addition, the latent heat of SSCPCM saturated with 90 wt% paraffin had a satisfactory latent heat capacity (178.3 kJ/kg) for latent heat thermal energy storage applications [66].

Figure 3.4 shows the DSC curve of the myristic acid. The phase change point of the PCM is observed to be 28.66/25.34 °C, with the latent heat of 61.778 J/g (melting process) and 58.810 J/g (freezing process). It indicates that the phase change temperature of myristic acid is suitable for building applications, however, its heat of fusion is not large enough for building energy efficiency utilization. In another study, Karaipekli *et al.* [67] reported the DSC analysis of capric-myristic acid/EG SSCPCM. The experimental results showed that the change in the phase change temperature had a little impact on the performance of the material, demonstrating good thermal stability.

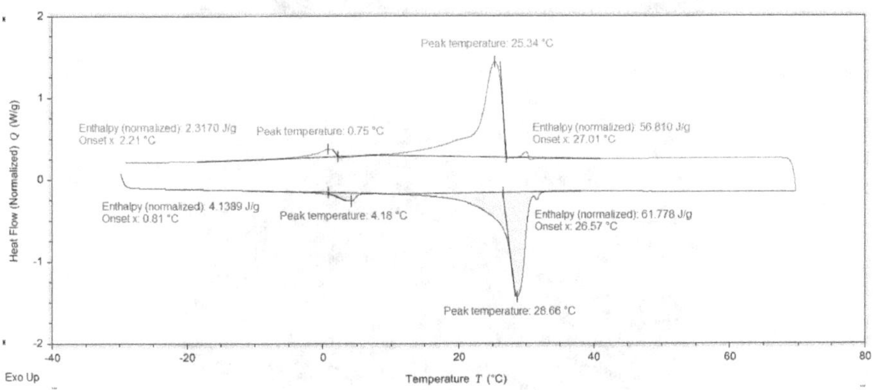

Figure 3.4 DSC curve of the myristic acid.

3.5.2 Morphology

The advantages of the SEM include:

- high magnification ratio which is continuously adjustable between 20 and 200,000 times

- large depth of field, wide field of view and rich stereoscopic image. Uneven surface microstructure can also be observed.
- test sample preparation is simple

Now a days, a X-ray spectrometer device is equipped with each SEM to simultaneously observe and analyze the microstructure morphology and micro area composition, thus, helping to analyze seepage, filling rate of the PCM, particle size distribution (PSD), etc. Figure 3.5 show the microstructure of EP. Figure 3.6 also shows a scanning

Figure 3.5 Scanning electron micrograph of EP.

Figure 3.6 Scanning electron micrograph of paraffin/EP CPCM.

electron micrograph of paraffin/EP CPCM. It can be observed that after complete adsorption of paraffin by EP, it still maintained its original stable form at the melting point, however, its microscopic appearance changed significantly. The porosity of the EP internal basic structure could be fully populated by paraffin, and the particles became dense with no visible porous structure of EP. When the external environmental temperature is higher than the melting point of paraffin, the composite particles would prevent the leakage of paraffin due to the capillary action and surface tension during the impregnation process.

In a representative study, Sari *et al.* [63] conducted the SEM analysis over shape stabilized paraffin/high density polyethylene CPCM. The experimental results indicated that paraffin was completely encapsulated by high density polyethylene, which played the role of support material, thus, completely avoiding the leakage of PCM. The authors observed that the maximum mass percentage of both paraffin types dispersed in the composites was 77 wt%. There was no leakage of paraffin up to this weight ratio even when it was melted [68].

3.5.3 Physical Stability

According to International Confederation for Thermal Analysis (ICTA), thermogravimetric analysis (TG) refers to a technique for measuring the relationship between a substance's mass and its temperature during program-controlled process. It is worth mentioning that the change is defined as the variation of quality rather than weight based on the effect of magnetic field. Strong magnetic materials have apparent weightlessness when reaching the Curie point, although no quality change is observed. TG analysis refers to the observation of material quality changes during the heating process.

Karaman *et al.* [69] carried out TG analysis to investigate the physical stability of polyethylene glycol/diatomite CPCM. The degradation of polyethylene glycol (PEG) was observed to start at 265 °C and ended at 510 °C, while the PEG confined in diatomite degraded in the temperature range of 245~500 °C before thermal cycling and 238~495 °C after thermal cycling, indicating physical stability during and after the thermal cycling process.

3.5.4 Chemical Stability

FT-IR spectroscopy is usually used in semiconductor manufacturing, among other numerous uses. It uses infrared spectrum to analyze impurity concentration by Fourier transform. It records absorption of Raman scattering, using photoconductance of solids, liquids and gases.

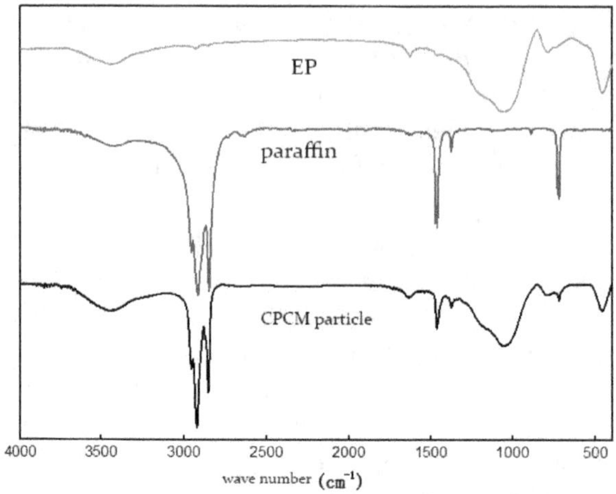

Figure 3.7 FT-IR spectra of EP, paraffin and paraffin/EP CPCM.

For instance, Figure 3.7 displays a FT-IR spectra for EP, paraffin and paraffin/EP CPCM particles. The shape of the composite spectrum revealed a superposition of EP and paraffin spectral curves, without obvious change in peak wave numbers, thus, showing that paraffin and EP are are physically mixed and the adsorbed paraffin did not react with EP.

Alkan *et al.* [70] carried out the FT-IR analysis on CPCM (n-eicosane microcapsules which were prepared by coating n-eicosane with polymethylmethacrylate (PMMA)). PMMA/eicosane microcapsules were observed to have close-grained and smooth surface indicating successful preparation and good chemical stability.

3.5.5 Adsorption Properties

BET test theory is based on the findings of Brunauer, Emmett and Taylor, who developed the model of multilayer adsorption and deduced the relationship between monolayer and multilayer adsorption quantity. BET equation is closer to the actual adsorption process of substances, thus, the test results are more accurate. A group of samples is tested under different nitrogen partial pressures for deducing multilayer adsorption quantity, which is linearly fitted by using BET equation in order to obtain a straight slope and intercept. This allows the calculation of specific surface area of the measured samples.

In a representative study, Sari and A. Karaipekli [66] conducted BET test on EG. In order to maximize the adsorption of the PCM (paraffin based) in EG, the authors also performed chemical treatment of EG. It was observed that EG had a specific area of 46 m^2/g after rapid expansion and exfoliation of expandable graphite in a furnace at 900 °C for 60s [66].

3.6 Properties of SSCPCMWs

Common building materials include wood, bamboo, stone, cement, concrete, metal, brick and tile, ceramics, glass, engineering plastics, composite materials, etc. It is difficult for the traditional building materials to meet the requirements of today's environmentally friendly, high efficiency and energy-saving buildings for sustainable development. Therefore, the combination of PCMs and building materials is conducive towards meeting the energy efficiency and energy saving for buildings.

SSCPCMs combined with a building maintenance structure are mainly in form of shape stabilized composite phase change material wallboard (SSCPCMW), usually made from the CPCMs and support materials. With CPCMs combined with building envelope structure, phase change can take place within the required temperature range, thus, offsetting the outdoor air temperature impact on the building envelope structure and reducing the indoor air temperature fluctuation.

Zhou *et al.* [71] studied SSCPCMWs and analyzed the thermal performance of the southward direct heat gain chamber installed with SSCPCMW by utilizing the enthalpy model. Experimental results concluded the influence of the following factors on the indoor air temperature: thermal physical properties of SSCPCMW, inner surface convective heat transfer coefficient, position and thickness of SSCPCMW installed on the wall structure (external thermal insulation wall plate material), etc. It was observed that: (1) under the current conditions, the optimum melting temperature was about 20 °C with the melting heat no less than 90 kJ/kg; (2) internal surface convection was the main factor limiting the heat storage capacity of SSCPCMW; (3) the heat storage effect of SSCPCMW should be better on the inner surface of the inner wall (south wall); (4) external wall insulation can significantly influence the operation effect and service cycle of SSCPCMW, as well as indoor temperature in winter; (5) SSCPCMW absorbed and stored solar energy during the day, and released heat at night, thus, improving indoor thermal comfort after sunset; (6) the SSCPCMW could make the lightweight envelope structure exhibit a heavyweight structure effect, uplifting the indoor temperature no less than 3 °C during the night [71].

3.7 Properties of Buildings Incorporated with SSCPCMWs

Shilei *et al.* [72] combined gypsum with the PCM to prepare SSCPCMW. The authors also established full-size experimental room and a full-size comparison facility. The phase change temperature and enthalpy of the PCM were measured by DSC. It was observed that DSC could effectively predict the thermal performance of the SSCPCMW in a full-size experimental room. Compared with full-size reference room, the experiment room installed with SSCPCMW could significantly reduce the power consumption of air conditioning system, and, at the same time, could also alleviate grid peak of electricity demand, thus, removing the mismatch between the supply and demand.

It is difficult to identify a single PCM which meets all of the ideal requirements in the practical situations. Therefore, optimum phase temperature and latent heat of PCMs are considered, and the comprehensive factors that influence the application are taken into consideration. Shape stabilized phase change material wallboard is a promising aspect for building energy efficiency and saving. Though there are still shortcomings for its applications, many significant achievements have been gained in the past decades. In the future, it is expected to play an important role in building construction.

3.8 Conclusions

A phase change material (PCM) refers to a substance that changes the state of matter within a certain temperature range and provides latent heat. The phase change process is accompanied by the release or absorption of a large extent of latent heat. By combining PCMs with building envelopes, a large amount of latent heat can be absorbed or released to offset the temperature change of the outdoor climate, thereby, effectively increasing the thermal inertia of the building envelope and reducing the indoor temperature fluctuations. PCMs have a significant effect on energy conservation and emission reduction, thus, improving building energy efficiency as well as residents' living comfort and work efficiency. These characteristics of composite phase change materials (CPCMs) formed by the combination of PCMs and building materials signify large application potential in construction field. In this respect, the current chapter briefly describes the background, classification, material properties, fabrication methods and application of PCMs from energy-saving aspects.

References

1. Perez-Lombard, L., Ortiz, J., and Pout, C. (2008) A review on buildings energy consumption information. *Energy and Buildings*, **40**(3), 394-398.
2. Zhang, H., and Yugong, X. U. (1996) *Application of Phase Change Material (PCM) in Building Envelopes,* Hefei of University of Science and Technology of China, China.
3. He, J., Qi-guang, W., Yong, Z., and Ge-cheng, Y. (2006) The progresses in preparation processes for porous aluminum. *Journal of Guangdong Non-Ferrous Metals*, **1**, 46-50.
4. Banhart, J. (2001) Manufacture, characterization and application of cellular metals and metal foams. *Progress in Materials Science*, **46**, 559-632.
5. Degischer, H. P. (1998) Innovative light metals: metal matrix composites and foamed aluminum. *Materials and Design*, **18**, 221-226.
6. Guo-tao, H., Cheng-dong, K., Qiang, Y., and Yu, D. (2012) Fabrication and properties of micron cell porous Al. *The Chinese Journal of Nonferrous Metals*, **8**, 2340-2346.
7. Ming-jun, Z., De-ping, H., and Feng, C. (2001) Compressive stress-strain behavior and energy absorption capability of porous aluminum alloy. *The Chinese Journal of Nonferrous Metals*, **S2**, 81-85.
8. Yang, L. (2009) Review on the architectural application of phase change material. *Architectural*, **10**, 65-67.
9. Zhang, Z. (2009) Building Energy Saving Application of Phase Change Material, p. 25-26.
10. Kaiming, D. (2009) *Preparation and Experimental Study of Thermal Insulating Wall Materials with PCM*, Chongqing University, China.
11. Zhu, M., Xiang, H., and Xu, W. (2013) A Melting Spinning Method for Preparing Intelligent Temperature-Regulating Fibers, patent CN 103243406 A.
12. Chen, C., Wang, X., and Shang, J. (2006) Numerical analysis and experimental study of heat change and discharge characteristics of middle temperature latent heat thermal storage tank. *Journal of Beijing University of Technology*, **32**(11), 996-1001.
13. Su, L., Ding, J., and Lei, X. (2012) Phase change materials for building insulation applications. *Energy Storage Science and Technology*,

1(2), 112-115.
14. Zhang, M. (2004) *Synthesis and Properties Study on Polyethylene Glycol/Poly (vinyl alcohol) Polymeric Solid-solid Phase Change Materials*, Jilin University, China.
15. Hawes, D. W, Banu, D., and Feldman, D. (1989) Latent heat storage in concrete. *Solar Energy Materials*, **19**(3-5), 335-348.
16. Kong, X. (2013) *Study on the Performance of Phase Change Material Combined with Building Envelope for Cold Storage*, TianJin University, China.
17. Sun, H. (2012) *Preparation of Phase Change Material: Functionalized Carbon Nanotubes/Palmitic Acid Composite*. Beijing University of Technology, China.
18. Ge, W., Sun, D., and Wang, A. (2015) Review on research progress and application of phase change building materials. *New Chemical Materials*, **2**, 229-231.
19. Shen, X., and Zhang, R., (2006) Study progress and application of phase change energy storage materials. *Energy Conservation Technology*, **24**(5), 460-463.
20. Zhang, Y., and Hu, H. (1996) *Theory and Application of Phase Change Energy Storage*, Hefei Press of University of Science and Technology of China, China.
21. Wang, X., and Lv, E. (1996) Solar energy solid-solid phase change thermal storage. *New Energy*, **18**(6), 9.
22. Fu, J. L., Yu, S. S., Li, S. J., and Ren, H. Y. (2005) Availability of tertiary oilfield water resources in western qaidam basin. *Journal of Salt Lake Research*, **13**, 17-21.
23. Zou, F. (2006) *Research on Heat Transfer Property of Paraffin Phase Change Material*, PhD Thesis, Shanghai Maritime University, China.
24. Xu, C., Yang, Z., and Lin, S. (2010) *Petroleum Refining Engineering*, 4th edition, Petroleum Industry Press, China.
25. Sharma, A., Tyagi, V. V., Chen, C. R., and Buddhi, D. (2009) Review on thermal energy storage with phase change materials and applications. *Renewable and Sustainable Energy Reviews*, **13**(2), 318-345.
26. Kong, X., Zhong, Y., Rong, X., Min, C., and Qi, C. (2016) Building energy storage panel based on paraffin/expanded perlite: Preparation and thermal performance study. *Materials*, **9**(2), 70.
27. Dincer, I., and Rosen, M. A. (2002) *Thermal Energy Storage, Systems and Applications*, John Wiley & Sons, UK.
28. Cabeza, L. F., Mehling, H., Hiebler, S., and Ziegler, F. (2002) Heat transfer enhancement in water when used as PCM in thermal energy storage. *Applied Thermal Engineering*, **22**, 1141-1151.
29. Kenisarin, M. M. (2010) High-temperature phase change materials for thermal energy storage. *Renewable and Sustainable Energy Reviews*, **14**, 955-970.
30. Zalba, B., Marın, J. M., Cabeza, L. F., and Mehling, H. (2003) Review

on thermal energy storage with phase change: materials, heat transfer analysis and applications. *Applied Thermal Engineering*, **23**(3), 251-283.

31. Ling, S., Gao, X., and Yin, H. (2007) State-of-the-art of PCM for heat storage at low temperature. *Guangdong Chemical Industry*, **3**, 48-51.
32. Wu, L., Chen, W., and Qian, J. (2017) Thermophysical Property of organic- inorganic phase change mixed material. *Packaging Engineering*, **9**, 113-117.
33. Abhat, A. (1983) Low temperature latent heat thermal energy storage: heat storage materials. *Solar Energy*, **30**, 313-332.
34. Tyagi, V. V., and D. Buddhi (2007) PCM thermal storage in buildings: A state of art. *Renewable and Sustainable Energy Reviews*, **11**(6), 1146-1166.
35. Royon, L., Guiffant, G., and Flaud, P. (1997) Investigation of heat transfer in a polymeric phase change material for low level heat storage, *Energy Conversion and Management*, **38**, 517-524.
36. Heckenkamp, J., and Baumann, H. (1997) Latent wärme speicher. *Sonderdruck aus Nachrichten*, **11**, 1075-1081.
37. Naumann, R., and Emons, H. H. (1989) Results of thermal analysis for investigation of salt hydrates as latent heat-storage materials. *Journal of Thermal Analysis*, **35**, 1009-1031.
38. Belton, G., and Ajami, F. (1973) *Thermochemistry of Salt Hydrates*, Report No. NSF/RANN/SE/GI27976/TR/73/4, Pennsylvania University, USA.
39. Hawes, D. W., Feldman, D., and Banu, D. (1993) Latent heat storage in building materials, *Energy and Buildings*, **20**, 77-86.
40. Nagano, K., Mochida, T., Iwata, K., Hiroyoshi, H., and Domanski, R. (2000) Thermal Performance of $Mn(NO_3)_2{\cdot}6H_2O$ as a new PCM for Cooling System. *5th Workshop of the IEA ECES IA Annex*, Japan.
41. Lane, G. A. (1980) Low temperature heat storage with phase change materials. *International Journal of Ambient Energy*, **1**, 155-168.
42. Regin, A. F., Solanki, S. C., and Saini, J. S. (2008) Heat transfer characteristics of thermal energy storage systems using PCM capsules: a review. *Renewable and Sustainable Energy Reviews*, **12**(9), 2438-2458.
43. Cabeza, L. F., Castell, A., Barreneche, C., de Gracia, A., and Fernandez, A. I. (2001) Materials used as PCM in thermal energy storage in buildings: a review. *Renewable and Sustainable Energy Reviews*, **15**(3), 1675-1695.
44. Telkes, M. (1975) Thermal Storage for Solar Heating and Cooling. *Workshop on Solar Energy Storage Subsystems for the Heating and Cooling of Buildings*, USA.
45. Lindner, F. (1996) Wärme speicherung mit Salzen und Salzhydraten. *KI - Luft- und Kältetechnik*, **10**, 462-467.
46. Wada, T., Yokotani, F., and Matsuo, Y. (1984) Equilibria in the aque-

ous ternary system containing Na^+, CH_3CO^{2-}, and $P_2O_7^{4-}$ between 38 and 85°C. *Bulletin of the Chemical Society of Japan*, **57**, 1671-1672.
47. Mettawee, E. S. and Assassa, G. M. R. (2007) Thermal conductivity enhancement in a latent heat storage system. *Solar Energy*, **81**(7), 839-845.
48. Yang, S. (2009) Study on performance of nano-aluminum/paraffin wax composite phase change material. *Materials Review*, **24**, 20-22.
49. Li, T. X., Wu, D. L., He, F., and Wang, R. Z. (2017) Experimental investigation on copper foam/hydrated salt composite phase change material for thermal energy storage. *International Journal of Heat and Mass Transfer*, **115**, 148-157.
50. Zhang, H., Wang, N., Zhu, D., and Yang, S. (2011) Study on performance of nano-copper/paraffin wax composite phase change material. *Materials Review*, S1.
51. Zhang, Z., Wang, X., and Fang, X. (2006) Structure and thermal properties of composite paraffin /expanded graphite phase-change material. *Journal of South China University of Technology (Natural Science Edition)*, **3**, 1-5.
52. Zhao, J. (2008) Polyethylene glycol/expanded graphite phase change composites for thermal storage, *Modern Chemical Industry*, **9**, 46-47.
53. Wang, J., Xie, H., and Xin, Z. (2009) Thermal properties of paraffin based composites containing multi-walled carbon nanotubes. *Thermochimica Acta*, **488**(1), 39-42.
54. Cao, Y., and Guo, M. (2016) Graphene materials and its applications. *Petrochemical Technology*, **10**, 1149-1159.
55. Wang, X. (2013) Thermophysical properties of paraffin-based composite phase change material filled with carbon nanotubes. *Journal of Thermal Science and Technology*, **2**,124-130.
56. Zhang, D., Kang, W., and Li, K. (2007) Progress in studies of phase change composites, *Functional Materials*, **12**, 1936-1940.
57. Chen, Z., Ma, L., and Yu, F., (2010) The research progress of organic phase change materials and its composites. *New Chemical Materials*, **9**, 42-44.
58. Ren, J., Ma, B., Si., W., Zhou, X., and Li, C, (2014) Preparation and analysis of composite phase change material used in asphalt mixture by sol–gel method. *Construction and Building Materials*, **71**, 53-62.
59. Zhang, X. (2012) *Preparation and Properties of Stearic Acid/SiO_2 Composite Phase Change Material*, Harbin Institute of Technology, China.
60. Zhou X., Xiao, H., Feng, J., Zhang, C., and Jiang, Y. (2008) Research progress in thermal control phase change materials and composites preparation. *Materials Review*, **11**, 30-33.
61. Jiang, Y., Zhang, Y., and Zheng, S. (2011) Preparation of composite phase change materials and its application development. *China Non-*

metallic Mining Industry Herald, **03**, 4-7.

62. Huang, J. (2005) *Study on Composite Phase Change Energy Storage Materials by Molten Salt Spontaneous Infiltrating Porous ceramic Performs*, Guangdong University of Technology, China.
63. Li, H., Ji, Z., Wang, J., and Xin, Z. (2008) Research status of phase change composite materials. *Materials Review*, **S2**, 248-251.
64. Lafdi, K., Mesalhy, O., and Elgafy, A. (2008) Graphite foams infiltrated with phase change materials as alternative materials for space and terrestrial thermal energy storage applications. *Carbon*, **46**(1), 159-168.
65. Yao, C., Kong, X., Li, Y., Du, Y., and Qi, C. (2018) Numerical and experimental research of cold storage for a novel expanded perlite-based shape-stabilized phase change material wallboard used in building. *Energy Conversion and Management*, **155**, 20-31.
66. Sari, A., and Karaipekli, A. (2007) Thermal conductivity and latent heat thermal energy storage characteristics of paraffin/expanded graphite composite as phase change material. *Applied Thermal Engineering*, **27**(8), 1271-1277.
67. Karaipekli, A., and Sari, A. (2008) Capric-myristic acid/expanded perlite composite as form-stable phase change material for latent heat thermal energy storage. *Renewable Energy*, **33**(12), 2599-2605.
68. Sari, A. (2004) Form-stable paraffin/high density polyethylene composites as solid-liquid phase change material for thermal energy storage: preparation and thermal properties. *Energy Conversion and Management*, **45**(13), 2033-2042.
69. Karaman, S., Karaipekli, A., Sari, A., and Bicer, A. (2011) Polyethylene glycol (PEG)/diatomite composite as a novel form-stable phase change material for thermal energy storage. *Solar Energy Materials and Solar Cells*, **95**(7), 1647-1653.
70. Alkan, C., Sari, A., and Karaipekli, A. (2011) Preparation, thermal properties and thermal reliability of microencapsulated n-eicosane as novel phase change material for thermal energy storage. *Energy Conversion and Management*, **52**(1), 687-692.
71. Zhou, G., Zhang, Y., Lin, K., and Xiao, W. (2008) Thermal analysis of a direct-gain room with shape-stabilized PCM plates. *Renewable Energy*, **33**(6), 1228-1236.
72. Shilei, L. V., Guohui, F., Neng, Z., and Li, D. (2007) Experimental study and evaluation of latent heat storage in phase change materials wallboards. *Energy and Buildings*, **39**(10), 1088-1091.

4

Nanofluid and Phase Change Material Integrated into a Photovoltaic Thermal System

Mohammad Hosseinzadeh,[a,1] Mohammad Sardarabadi[b] and Mohammad Passandideh-Fard[a,2]

[a]*Department of Mechanical Engineering, Ferdowsi University of Mashhad, Mashhad, Iran*
[b]*Department of Energy, Quchan University of Technology, Quchan, Iran*

[1]*The author is currently a researcher at Research Institute of Food Science and Technology (RIFST), Mashhad, Iran*
[2]*Corresponding author:* mpfard@um.ac.ir

4.1 Introduction

Nowadays, due to the price and environmental effects of fusil fuels, the use of different solar systems has increased. A photovoltaic (PV) module directly converts the solar energy into electrical energy. Increasing the temperature of photovoltaic cells is a major limitation that reduces the efficiency of PV modules [1]. Moreover, this can damage the photovoltaic cells [2]. Hence, there is a growing intention to find useful methods for cooling the photovoltaic cells such as using fins [3], phase change materials (PCMs) [4], and thermoelectric modules [5]. A photovoltaic thermal (PVT) system consists of a PV module integrated with a cooling system such as thermal collectors. The schematic of a PVT system including a PV module and a sheet-and-tube collector is shown in Figure 4.1. In the PVT systems, the extracted energy from the PV module is utilized as a useful thermal energy. Hence, the PVT is a hybrid system to generate the electrical and thermal energies, simultaneously. The type of the coolant fluid dramatically affects the performance of PVT systems. Many studies in literature have investigated the effect of different nanofluids on the performance of different solar systems such as solar collectors [6]. Use of nanofluids as coolants of PVT systems enhances the heat removal performance of the system compared to conventional fluids [7,8]. However, this

Phase Change Materials, edited by Vikas Mittal

has some drawbacks such as high cost, limited time of stability and increased pressure drop in the collector [8,9]. It should be mentioned that nanofluids can also be used as the optical filter in the PVT systems [10]. This filter absorbs the solar energy that cannot be employed by the photovoltaic cells [11]. The evaluation of using nanofluids as the optical filter in PVT systems is not of interest of this chapter.

Figure 4.1 The schematic of a PVT system.

As mentioned above, an effective method to cool down the photovoltaic cells is the use of PCMs. Generally, the PCMs can be divided in three categories: organic (paraffins and non-paraffins), inorganic (salt hydrates and metallics) and eutectic (mixtures of organic-organic, inorganic-inorganic or organic-inorganic PCMs). The properties of different PCMs have been discussed thoroughly in the previous studies [12-14]. Several literature reports have studied the performance of PV modules with PCMs (referred as PV/PCM systems) [15-18]. In this chapter, the effect of using the PCMs in the PVT systems is investigated. As displayed in Figure 4.2, the PCMs can be used in the structure of PVT systems (known as PVT/PCM systems). Moreover, the encapsulated phase change material (referred as EPCM) slurry can be utilized as the coolant of PVT systems. In the EPCM slurry, the PCM cores are wrapped by shells and the capsules are dispersed in a carrier fluid. Based on the EPCMs size, these can be classified in three categories: nano-encapsulated phase change materials, micro-encapsulated phase change materials and macro-encapsulated phase change materials. Dispersion of EPCMs in a fluid enhances its latent heat storage capacity [19]. However, similar to nanofluids, there are drawbacks of using EPCM slurries in the PVT systems, which include high cost, limited time of stability and increased the pressure drop in the collector. Moreover, the EPCMs may be broken in the pumping circulations [20].

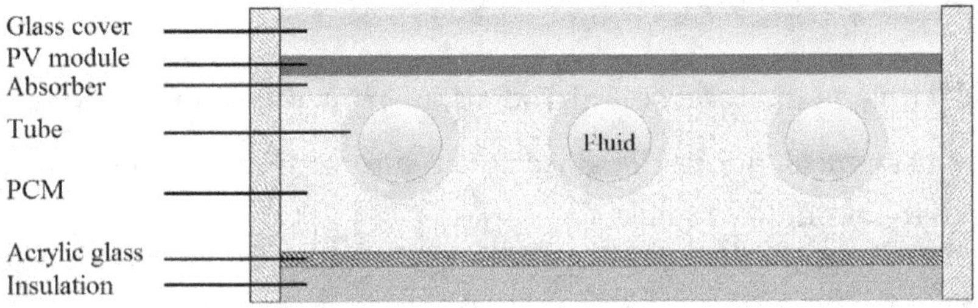

Figure 2. A schematic of a PVT/PCM system.

In this chapter, the thermodynamic analysis of different PVT systems is presented from the energy and exergy viewpoints. In order to investigate the performance of PVT systems, different approaches are studied and appropriate equations are expressed. Moreover, in this chapter, the effect of using nanofluids (as the coolant fluid) as well as PCMs on the performance of PVT systems is evaluated. Furthermore, the literature studies on this aspect are reviewed and the important parameters in the nanofluid-based PVT systems and PCM-based PVT systems are investigated.

4.2 Thermodynamic Analysis

The performance of PVT systems can be investigated using the first and second laws of thermodynamics (energy and exergy analyses). As mentioned before, in the PVT systems, the electrical and thermal energies are generated. The quality of the electrical and thermal energies is not the same: the electrical energy is equivalent to the available work, whereas a portion of the thermal energy can be employed as the available work. Contrary to the energy analysis, the exergy analysis considers the quality of energies. Therefore, the exergy analysis is necessary to evaluate the real performance of PVT systems.

4.2.1 Energy Analysis

In order to investigate the performance of a PVT system from the energy viewpoint, the PVT system is considered as a control volume.

Moreover, the system is assumed to be in semi-steady state condition. The absorbed solar irradiation by the PVT system can be calculated using the following equation:

$$\dot{E}_{sun} = \dot{G}\ A_c\ \tau_g\ \alpha_{cell} \tag{1}$$

in which $\dot{G}$ is the total solar irradiation, A_c is the collector area, τ_g is the transmissivity of the glass cover, and α_{cell} represents the absorptivity of photovoltaic cells.

The output thermal power of the PVT system is expressed as follows:

$$\dot{E}_{th} = \dot{m}_f\ C_{p,f}\ (T_{f,out} - T_{f,in}) \tag{2}$$

In the above equation, $\dot{m}_f$ refers to the mass flow rate of the coolant fluid, $C_{p,f}$ is the specific heat capacity of the coolant fluid, and $T_{f,in}$ and $T_{f,out}$ are the fluid temperature at the inlet and outlet of the collector, respectively. If the coolant fluid is a nanofluid, the specific heat capacity of the nanofluid can be calculated based on one of these equations [21]:

$$C_{p,nf} = \phi\, C_{p,np} + (1-\phi)\ C_{p,bf} \tag{3}$$

$$C_{p,nf} = \frac{\phi\ (\rho_{np} C_{p,np}) + (1-\phi)\ (\rho_{bf} C_{p,bf})}{\rho_{nf}} \tag{4}$$

where $C_{p,np}$ and $C_{p,bf}$ are the specific heat capacity of nanoparticles and base fluid, respectively. Moreover, ρ_{np}, ρ_{bf} and ρ_{nf} refer to the density of nanoparticles, base fluid and nanofluid, respectively. ϕ is the volume fraction of nanoparticles, which is defined by the following equation:

$$\phi = \frac{m_{np} / \rho_{np}}{m_{np} / \rho_{np} + m_{bf} / \rho_{bf}} \tag{5}$$

In the above equation, m_{np} and m_{bf} are the mass of nanoparticles and base fluid, respectively. In equation (4), the density of nanofluid is calculated using the physical principle of two phase mixture:

$$\rho_{nf} = \phi \, \rho_{np} + (1-\phi) \, \rho_{bf} \tag{6}$$

It should be mentioned that based on the previous research [21-23], equation (4) has better agreement with the experimental data compared to Equation (3).

The output electrical power of the PVT system is expressed as follows [24,25]:

$$\dot{E}_{el} = V_{oc} \, I_{sc} \, FF \tag{7}$$

in which V_{oc} is the open circuit voltage, I_{sc} is the short circuit current, and FF represents the fill factor [24,26]. As mentioned before, the quality of the electrical and thermal energies is different. In order to evaluate the performance of the PVT system based on the thermal energy, the output electrical power can be converted to an equivalent thermal power as [27,28]:

$$\dot{E}_{el,th} = \frac{\dot{E}_{el}}{C_f} \tag{8}$$

where C_f is a conversion factor that can be considered about 0.38 [27-29].

Based on the cooling system, PVT systems can be divided into two categories: passive (thermosyphon) and active (with pump). A schematic of an active PVT experimental setup is displayed in Figure 4.3. In the active PVT system, the consumed electrical power by the pump can be calculated as follows [30,31]:

$$\dot{E}_p = \frac{\dot{m}_f \, \Delta P}{\rho_f \, \eta_p} \tag{9}$$

In the above equation, ΔP is the pressure drop of the coolant fluid in the collector, and η_p is the pump efficiency.

In the passive PVT systems, the electrical energy efficiency of the system is expressed as follows [32]:

$$\eta_{el} = \frac{\dot{E}_{el}}{\dot{E}_{sun}} \tag{10}$$

In the active PVT systems, the electrical energy efficiency of the system can be calculated using equation (11) [7,30,31,33]. However, in some studies, equation (10) has been used for the active PVT system. Therefore, this matter should be considered in future research efforts.

$$\eta_{el} = \frac{\dot{E}_{el} - \dot{E}_{P}}{\dot{E}_{sun}} \tag{11}$$

In the numerical studies of PVT systems, the electrical energy efficiency can be calculated as follows [2,34-36]. In this equation, the electrical energy efficiency depends on the photovoltaic cells temperature:

$$\eta_{PV} = \eta_r \, [1 - \beta \, (T_{cell} - T_r)] \tag{12}$$

$$\eta_{el} = PF \, \eta_{PV} \tag{13}$$

where η_{PV} is the electrical efficiency of the PV module, η_r is the reference efficiency of the PV module, and β is the temperature coefficient of the PV module. T_{cell} and T_r are the photovoltaic cells temperature and reference temperature, respectively. PF refers to the packing factor of the PVT system that is defined as the PV module area to the collector area (A_{pv} / A_c).

The thermal and overall energy efficiencies of the PVT system are expressed as follows [1,2,7]:

$$\eta_{th} = \frac{\dot{E}_{th}}{\dot{E}_{sun}} \tag{14}$$

$$\eta_{ov} = \eta_{el} + \eta_{th} \tag{15}$$

4.2.2 Exergy Analysis

Similar to the energy analysis, in the exergy analysis, the PVT system is considered as a control volume. Moreover, the system is assumed to be in semi-steady state condition. The exergy of the absorbed solar irradiation by the PVT system can be calculated using the following equations [32,37-39]:

$$\dot{E}x_{sun} = \dot{E}_{sun}\,(1-\frac{T_{amb}}{T_{sun}}) \tag{16}$$

$$\dot{E}x_{sun} = \dot{E}_{sun}\,(1-\frac{4T_{amb}}{3T_{sun}}) \tag{17}$$

$$\dot{E}x_{sun} = \dot{E}_{sun}\,[1-\frac{4T_{amb}}{3T_{sun}}+\frac{1}{3}(\frac{T_{amb}}{T_{sun}})^4] \tag{18}$$

In the above equations, T_{amb} represents the ambient temperature, and T_{sun} is the sun temperature as a blackbody. Chow *et al.* [32] demonstrated that the difference between equations (16)-(18) is less that 2%. However, equation (18) has been more frequently used in the literature as compared to other two equations [37].

The output thermal exergy of the PVT system can be calculated using Equations (19) or (20) [38,40,41]:

$$\dot{E}x_{th} = \dot{E}_{th}\,(1-\frac{T_{amb}}{T_{f,out}}) \tag{19}$$

$$\dot{E}x_{th} = \dot{E}x_{mass,out} - \dot{E}x_{mass,in} = \dot{m}_f\,\left[(h_{f,out}-h_{f,in})-T_{amb}(s_{f,out}-s_{f,in})\right]\rightarrow$$

$$\dot{E}x_{th} = \dot{m}_f\,C_{p,f}\left[(T_{f,out}-T_{f,in})-T_{amb}\ln(\frac{T_{f,out}}{T_{f,in}})\right] \tag{20}$$

where $\dot{E}x_{mass,in}$ and $\dot{E}x_{mass,out}$ represent the flow (stream) exergy at the inlet and outlet of the collector, respectively. Moreover, $h_{f,in}$ and $h_{f,out}$ are the fluid enthalpy at the inlet and outlet of the collector, respectively. Furthermore, $s_{f,in}$ and $s_{f,out}$ are the fluid entropy at the inlet and outlet of the collector, respectively.

It should be mentioned that, in equation (19), the thermal exergy is derived based on the Carnot efficiency: the thermal exergy is an available work extracted from the thermal power at the outlet temperature of the coolant fluid (as the heat source temperature). On the other hand, in equation (20), the thermal exergy is calculated by the difference between the flow exergy at the inlet and outlet of the collector.

As mentioned before, the electrical energy is equivalent to the available work. Thus, the output electrical exergy of the PVT system is equal to the electrical power. Therefore, in the passive PVT systems, the output electrical exergy of the system is defined as follows [32]:

$$\dot{E}x_{el} = \dot{E}_{el} \tag{21}$$

As shown in equation (22), in the active PVT systems, the output electrical exergy of the system is defined as the difference between the output electrical power and the consumed pumping power [7,30,31,33]. However, in some research studies, this matter was not considered and the output electrical exergy was calculated based on equation (21).

$$\dot{E}x_{el} = \dot{E}_{el} - \dot{E}_{p} \tag{22}$$

The exergy destructions of the PVT system as well as the entropy generation of the system are calculated using equations (23) and (24), respectively [34,42,43].

$$\dot{E}x_{dest} = \dot{E}x_{sun} - \dot{E}x_{el} - \dot{E}x_{th} \tag{23}$$

$$\dot{S}_{gen} = \frac{\dot{E}x_{dest}}{T_{amb}} \tag{24}$$

The electrical, thermal and overall exergy efficiencies of the PVT system are expressed as follows [7,33]:

$$\varepsilon_{el} = \frac{\dot{E}x_{el}}{\dot{E}x_{sun}} \tag{25}$$

$$\varepsilon_{th} = \frac{\dot{E}x_{th}}{\dot{E}x_{sun}} \tag{26}$$

$$\varepsilon_{ov} = \varepsilon_{el} + \varepsilon_{th} \tag{27}$$

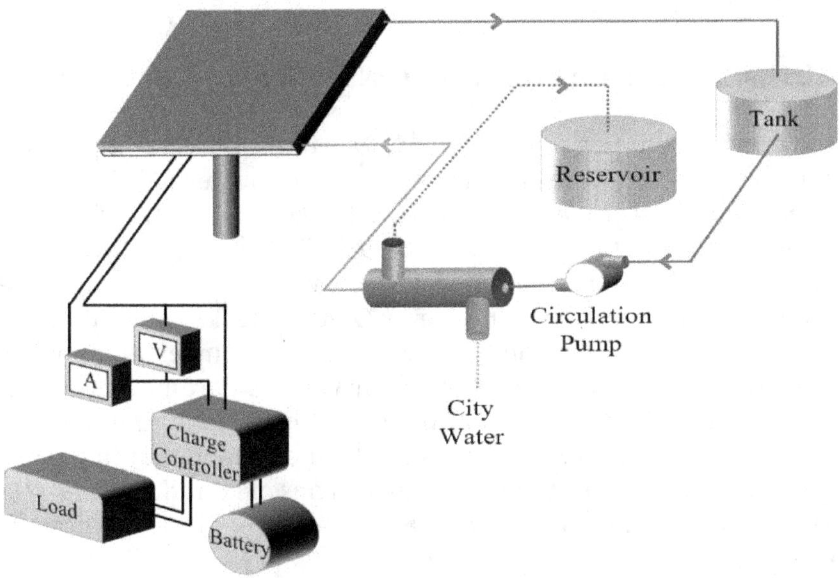

Figure 4.3 Schematic of an active PVT experimental setup.

4.3 Nanofluid-based PVT Systems

In this section, the research studies regarding nanofluid-based PVT systems are reviewed and the studied parameters are evaluated. The investigated parameters in the previous studies on nanofluid-based PVT systems include: thermophysical properties of nanofluids [7,25,35,44-54], mass/volume fraction of nanoparticles [7,25,35,40,49,51,52,55-67], flow regime in the cooling system [7,40,49], base fluids [7,46,68], particle size [7,50,62], nanoparticle type [7,25,27,35,44,46,48,52,66,68-71], mass/volumetric flow rate of nanofluids in the cooling system [25,40,44,51,54,56,59,60,64,65,67,68,70,71], Reynolds number

[7,25,56,62,66,68], nanofluids temperature at the inlet of the cooling system [36,51,65], external magnetic field [55,57], solar irradiation [36,44,48,54,55,63,65,70,72], ambient temperature [65,67,72], wind speed [65,67,72] and glass cover [7]. In this chapter, the most important parameters such as thermophysical properties of nanofluids, mass/volume fraction of nanoparticles, flow regime, base fluids and particle size have been reviewed. It should be mentioned that for the rest of the chapter, the nanofluid-based PVT systems are referred to as NPVT systems.

4.3.1 Thermophysical Properties of Nanofluids

In the previous studies regarding NPVT systems, the thermophysical properties of silver-water (Ag-water), alumina-water (Al_2O_3-water), alumina-water/ethylene glycol mixture (Al_2O_3-water/EG), silica-water (SiO_2-water), titanium dioxide-water (TiO_2-water), silicon carbide-water (SiC-water), copper-water (Cu-water), copper oxide-water (CuO-water) and zinc oxide-water (ZnO-water) nanofluids were numerically and experimentally evaluated. The investigations on nanofluid thermophysical properties for various parameters are reported in Table 4.1. As shown in the table, the base fluid in most of the reported studies is water. It should be mentioned that in the numerical studies, the thermophysical properties of nanofluids were estimated with different relations that are available in the literature [21,22].

The literature review reveals that the effects of mass/volume fraction of nanoparticles and nanofluid temperature on the thermophysical properties of nanofluids were extensively investigated, whereas only a few studies evaluated the effect of other parameters such as particle size and shear stress on the thermophysical properties of nanofluids in the NPVT systems [7,45,50]. Based on the available studies [35,44,47-49], the density, thermal conductivity and viscosity of nanofluids are increased by raising the mass/volume fraction of nanoparticles. However, the specific heat capacity of nanofluids is decreased. The experimental investigations [47] demonstrated that using SiC-water nanofluid with a mass fraction of 3% (3 wt%) increases the density, thermal conductivity and viscosity of the fluid by 0.0082%, 8.3% and 0.18% compared to those of pure water, respectively. Moreover, according to the experimental studies [48], for a same volume fraction of nanoparticles, the thermal conductivity and viscosity of SiC-water nanofluid are higher than those of Al_2O_3-water and CuO-water nanofluids. However, the density of CuO-water nanofluid is higher than that of two other nanofluids (SiC-water and Al_2O_3-water).

Table 4.1 Numerical and experimental studies of the effects of various parameters on thermophysical properties in NPVT systems

Nanofluid	Parameters whose effects on thermophysical properties were investigated	
	Numerical studies	Experimental studies
Ag-water	Volume fraction [35]	N/A
Al_2O_3-water	Mass/Volume fraction [7,25,35,46,52]	Volume fraction [48] Nanofluid temperature [48]
Al_2O_3-water/EG	Volume fraction [7]	N/A
SiO_2-water	Mass fraction [52]	Mass/Volume fraction [44,50] Nanofluid temperature [44,50] Particle size [50]
TiO_2-water	Mass/Volume fraction [7,25,49,52] Particle size [7]	Mass fraction [44] Nanofluid temperature [44]
SiC-water	N/A	Mass/Volume fraction [44,47,48,53,54] Nanofluid temperature [44,47,48,53,54]
Cu-water	Mass/Volume fraction [46,51] Nanofluid temperature [51]	N/A
CuO-water	N/A	Volume fraction [48] Nanofluid temperature [45,48] Shear stress [45]
ZnO-water	Mass fraction [52]	N/A

Despite the reduction of the density and viscosity of nanofluids, the thermal conductivity of nanofluids is enhanced by increasing the nanofluid temperature [44,53,54]. Enhancement of the thermal conductivity of nanofluids can be attributed to increasing the Brownian motion of nanoparticles [44,48]. Moreover, by increasing the particle size in a constant volume fraction of nanoparticles, the thermal conductivity and viscosity of nanofluids were reported to decrease [7,50].

4.3.2 Mass/volume Fraction of Nanoparticles

The mass/volume fraction of nanoparticles is an important parameter extensively investigated in the previous studies on NPVT systems.

As reported in Table 4.2, several research studies numerically and experimentally investigated the effect of mass/volume fraction of nanoparticles in various nanofluids including Ag-water, Al_2O_3-water, Al_2O_3-water/EG, SiO_2-water, TiO_2-water, SiC-water, Cu-water, ZnO-water, magnetite-water (Fe_3O_4-water), carbon nanotube-water (CNT-water), multi-walled carbon nanotube-water (MWCNT-water) and boehmite-water (AlOOH-water). Moreover, the investigated parameters of NPVT systems for the variation of mass/volume fraction of nanoparticles are mentioned in Table 4.2. It should be noted that since the density of nanoparticles is different, comparison of the performance of different nanoparticles with same volume fraction is not accurate; the comparison should be rather performed with same mass fraction. However, for a single nanoparticle type, the effect of various volume fractions on the performance of a NPVT system can be examined.

The Reynolds and Prandtl numbers of nanofluids can be calculated using the following equations:

$$\mathrm{Re}_{nf} = \frac{4\dot{m}_{nf}}{\pi d \mu_{nf}} \tag{28}$$

$$\mathrm{Pr}_{nf} = \frac{\mu_{nf} Cp_{nf}}{k_{nf}} \tag{29}$$

As mentioned in Section 4.3.1, the viscosity of nanofluids is increased by raising the mass/volume fraction of nanoparticles. Therefore, using equation (28), for a constant mass flow rate, the Reynolds number is decreased by increasing the mass/volume fraction of nanoparticles [49,52]. As discussed before, the thermal conductivity of nanofluids is also enhanced by increasing the mass/volume fraction of nanoparticles. However, the specific heat capacity of nanofluids is reduced. Thus, based on equation (29), the variation of the Prandtl number by increasing the mass/volume fraction of nanoparticles is related to the exact values of thermophysical properties of nanofluids (viscosity, thermal conductivity and specific heat capacity). In the numerical studies of NPVT systems, in order to determine the Prandtl number correctly, use of accurate relations for the thermophysical properties of nanofluids is important. The available literature on numerical studies [7,49] illustrated that the Prandtl number of Al_2O_3-water and TiO_2-water nanofluids is increased by raising the volume fraction of nanoparticles. However, another numerical research [52]

Table 4.2 The numerical and experimental investigations of the effect of mass/volume fraction of nanoparticles in the NPVT systems

Nanofluid	Numerical investigations	Experimental investigations	The investigated parameter for variation of mass/volume fraction of nanoparticles
Ag-water	[35]	[40]	The heat transfer coefficient; the nanofluid temperature at the outlet of cooling system; the absorber plate temperature; the output electrical power; the electrical, thermal, and overall energy efficiencies; the overall exergy efficiency.
Al_2O_3-water	[7,35,52,62,66]	[25]	The Reynolds number; the Prandtl number; the Nusselt number; the heat transfer coefficient; the friction factor; the pressure drop; the nanofluid temperature at the outlet of cooling system; the absorber plate temperature; the cells temperature; the output electrical power; the pumping power; the electrical, thermal, and overall energy efficiencies; the electrical, thermal, and overall exergy efficiencies; the frictional and thermal entropy generations.
Al_2O_3-water/EG	[7]	N/A	The Reynolds number; the Prandtl number; the Nusselt number; the heat transfer coefficient; the nanofluid temperature at the outlet of cooling system; the cells temperature; the electrical, thermal, and overall energy efficiencies; the electrical, thermal, and overall exergy efficiencies.
SiO_2-water	[52]	[64]	The Reynolds number; the Prandtl number; the Nusselt number; the heat transfer coefficient; the pressure drop; the nanofluid temperature at the outlet of cooling system; the cells temperature; the output electrical, thermal , and overall exergies; the electrical, thermal, and overall energy efficiencies; the electrical, thermal, and overall exergy efficiencies; the frictional and thermal entropy generations.
TiO_2-water	[7,49,52]	[25]	The Reynolds number; the Prandtl number; the Nusselt number; the heat transfer coefficient; the friction factor; the pressure drop; the nanofluid temperature at the outlet of cooling system; the cells temperature; the output electrical power; the pumping power; the electrical, thermal, and overall energy efficiencies; the electrical, thermal, and overall exergy efficiencies; the frictional and thermal entropy generations.
SiC-water	[66]	N/A	The cells temperature; the output electrical power; the pumping power; the electrical and thermal energy efficiencies.

Table 4.2 (continued)

Nanofluid	Numerical investigations	Experimental investigations	The investigated parameter for variation of mass/volume fraction of nanoparticles
Cu-water	[51]	N/A	The electrical, thermal, and overall energy efficiencies; the overall exergy efficiency.
ZnO-water	[52,65]	N/A	The Reynolds number; the Prandtl number; the Nusselt number; the heat transfer coefficient; the pressure drop; the nanofluid temperature at the outlet of cooling system; the cells temperature; the electrical, thermal, and overall energy efficiencies; the electrical, thermal, and overall exergy efficiencies; the frictional and thermal entropy generations.
Fe_3O_4-water	N/A	[55,57]	The nanofluid temperature at the cooling system; the cells temperature; the output electrical and thermal powers; the output overall exergy; the electrical and thermal energy efficiencies.
CNT-water (MWCNT-water)	[60,63,67]	[63]	The Nusselt number; the nanofluid temperature at the outlet of cooling system; the cells temperature; the output electrical and thermal powers; the electrical, thermal, and overall energy efficiencies.
AlOOH-water	N/A	[56]	The cells temperature; the nanofluid temperature at the cooling system; the output electrical power.

displayed a contrary trend for the Prandtl number. According to equation (29), the difference in the variations of the Prandtl number may be due to the precision used in the estimation of the viscosity and thermal conductivity of nanofluids in the mentioned literature studies.

Based on the previous studies [7,49,51,52,65], in the laminar flow regime, the electrical and thermal energy as well as exergy efficiencies of NPVT systems are improved by increasing the mass/volume fraction of nanoparticles. However, the effect of variation of mass/volume fraction of nanoparticles on the electrical energy and exergy efficiencies of NPVT systems is not significant. The experimental investigations [63] showed that the electrical, thermal and overall energy efficiencies of MWCNT-water nanofluid with 1 wt% are enhanced by 0.14%, 3.67% and 3.81% compared to those of pure water, respectively. In the laminar flow regime, the total entropy generation is decreased by increasing the mass/volume fraction of nanoparticles [52]. Comparison of the NPVT systems with Al_2O_3-water, SiO_2-water, TiO_2-water and ZnO-water nanofluids demonstrates that with same mass fraction of nanoparticles, SiO_2-water and Al_2O_3-water nanofluids are observed to have the highest and lowest total entropy generation as compared to other mentioned nanofluids, respectively [52].

4.3.3 Flow Regime

The flow regime in the cooling system of NPVT systems can be either laminar or turbulent. Based on the literature review, most of investigations on the NPVT systems were performed in the laminar regime [7]. However, a few studies evaluated the effect of a wider flow regime on the performance of NPVT systems [7,40,49]. As discussed before, in a constant mass flow rate, the Reynolds number is decreased by raising the volume fraction of nanoparticles. However, the numerical investigations [7,49] revealed that the Prandtl number of Al_2O_3-water and TiO_2-water nanofluids is increased. Hence, for Al_2O_3-water and TiO_2-water nanofluids, by increasing the volume fraction of nanoparticles, the trends of the Reynolds number and the Prandtl number are contradictory. The Nusselt number depends on the Reynolds number and the Prandtl number [7]. According to the previous studies [7,49], the effect of Reynolds number on the Nusselt number is more than the Prandtl number in the turbulent regime. However, in laminar regime, the effect of the Prandtl number is more pronounced.

Therefore, in the laminar regime, the Nusselt number is increased by raising the volume fraction of nanoparticles. However, the Nusselt number is decreased in the turbulent regime. It should be mentioned that another numerical study [52] reported an opposite trend for the variation of Nusselt number of Al_2O_3-water and TiO_2-water nanofluids by raising the mass fraction of nanoparticles in the laminar regime. This can be due to the contradictory trend of the Prandtl number (see Section 4.3.2). The convection heat transfer coefficient between the nanofluid and the cooling system can be calculated as follows:

$$h = \frac{Nu_{nf} k_{nf}}{d} \tag{30}$$

As mentioned before, in the laminar flow regime, the Nusselt number and the thermal conductivity of Al_2O_3-water and TiO_2-water nanofluids are increased by raising the volume fraction of nanoparticles. Therefore, using equation (30), the convection heat transfer coefficient is increased. Moreover, as discussed above, the Nusselt number is decreased by increasing the volume fraction of nanoparticles in the turbulent regime. Hence, the variations of the Nusselt number and the thermal conductivity of nanofluids are opposite of each other. Thus, based on equation (30), the variation of the convection heat transfer coefficient by increasing the volume fraction of nanoparticles depends on the exact value of the thermal conductivity of nanofluids. Therefore, in numerical studies, use of accurate relations is of paramount importance to estimate the thermal conductivity of nanofluids. The previous studies [7,49] illustrated that in the turbulent regime, the convection heat transfer coefficient of Al_2O_3-water and TiO_2-water nanofluids are decreased by increasing the volume fraction of nanoparticles. Therefore, the photovoltaic cells temperature is increased by raising the volume fraction of nanoparticles in the turbulent regime. However, in the laminar regime, the photovoltaic cells temperature is decreased. Consequently, the electrical energy and exergy efficiencies of NPVT systems with Al_2O_3-water and TiO_2-water nanofluids are reduced by increasing the volume fraction of nanoparticles in the turbulent regime. However, the corresponding efficiencies are enhanced in the laminar regime. It should be mentioned that the experimental investigations [40] on Ag-water nanofluid demonstrated that increasing the mass fraction of nanoparticles enhances the electrical energy efficiency of NPVT system in the turbulent regime. This is in contrast to the results of [7,49] for Al_2O_3-water and TiO_2-water nanofluids. Increase in electrical energy efficiency of NPVT system can be attributed to the enhancement of convection heat transfer of Ag-water nanofluid, based on equation (30). Therefore, it seems that more numerical and experimental investigations are necessary in order to evaluate the effect of mass/volume fraction of nanoparticles in the turbulent regime for various nanofluids in the NPVT systems.

4.3.4 Base Fluids

Literature survey reveals that most of previous studies regarding the NPVT systems used water as the base fluid. Only a few studies have investigated the effect of other base fluids such as ethylene glycol [46,68], water/EG mixture with 40:60 wt% [7] and glycerin [68] on the performance of NPVT systems. The previous studies [7] showed that the viscosity of Al_2O_3-water/EG nanofluid is higher than that of Al_2O_3-water nanofluid with same volume fraction of nanoparticles. Thus, based on equation (28), in a constant mass flow rate, the Reynolds number of Al_2O_3-water/EG nanofluid is lower than that of Al_2O_3-water nanofluid. However, the results of thermophysical properties of nanofluids demonstrated that the Prandtl number of Al_2O_3-water/EG nanofluid is higher than that of Al_2O_3-water nanofluid. Therefore, based on the numerical results [7], in the turbulent regime, the Nusselt number and convection heat transfer coefficient of Al_2O_3-water nanofluid are higher than those of Al_2O_3-water/EG nanofluid. Thus, the photovoltaic cell temperature of Al_2O_3-water nanofluid is lower than that of Al_2O_3-water/EG nanofluid. Consequently, the electrical energy and exergy efficiencies of the NPVT system with Al_2O_3-water nanofluid are higher than those of Al_2O_3-water/EG nanofluid. Moreover, in another numerical study [68], the effect of the base fluids including water, ethylene glycol and glycerin as well as the nanoparticles of SiC, CuO and Al_2O_3 on the Nusselt number and pressure drop of NPVT systems was investigated. This study illustrated that in the turbulent regime, the SiC-based and water-based nanofluids have the highest Nusselt number compared to other considered nanofluids with a constant Reynolds number. Moreover, the water-based and glycerin-based nanofluids have the lowest and highest pressure drop in the NPVT systems, respectively. In addition, the Nusselt number and pressure drop of nanofluids are increased by raising the Reynolds number.

4.3.5 Particle Size

As mentioned before, with a constant mass/volume fraction of nanoparticles, increasing the particle size decreases the thermal conductivity and viscosity of nanofluids [7,50]. Therefore, at a constant mass flow rate, based on equation (28), the Reynolds number is increased by raising the particle size. However, the available studies [7] demonstrated that using equation (29), the Prandtl number of TiO_2-water nanofluid is decreased. Thus, in the laminar flow regime, the Nusselt number and convection heat transfer coefficient of nanofluid are reduced by increasing the particle size. However, the Nusselt number and convection heat transfer coefficient of nanofluid are enhanced in turbulent regime. The numerical results [7] showed that in the laminar regime, the total energy and exergy efficiencies of NPVT system are slightly reduced by increasing the particle size. However, in the turbulent regime, the corresponding efficiencies are slightly enhanced. Moreover, another numerical study [62] illustrated that at a constant volume fraction of nanoparticles, the Nusselt number and convection heat transfer coefficient of Al_2O_3-water nanofluid are decreased by increasing the particle size in the laminar regime. In addition, the effect of particle size on the convection heat transfer coefficient is increased by raising the volume fraction of nanoparticles. It should be mentioned that the pressure drop of Al_2O_3-water nanofluid is reduced by increasing the particle size; this may be attributed to the reduction of the nanofluid viscosity. As a result, in the numerical studies on NPVT systems, it is important to use accurate relations that consider the particle size, in order to estimate the thermophysical properties of nanofluids. Moreover, more numerical and experimental investigations are required to investigate the effects of particle size in the turbulent regime for various nanofluids in the NPVT systems.

4.4 PCM-based PVT Systems

In this section, the effect of using PCMs in the PVT systems is investigated. As mentioned before, the PCMs can be used in the structure of the PVT system (PVT/PCM system). Moreover, the micro-encapsulated phase change material (MEPCM) slurry can be utilized as the coolant of the PVT system. In this section, the previous studies regarding the PVT/PCM systems as well as the PVT systems with MEPCM slurries are also reviewed.

4.4.1 PVT/PCM Systems

The investigated parameters regarding the PVT/PCM systems in the previous studies include the PCM type [73,74], melting temperature of PCM [75,76], thermal conductivity of PCM [73,75,77,78], type of coolant in the PVT/PCM system [79,80], position of PCM in the PVT/PCM system [81], thickness of PCM [74-76,81], PCM mass [82], mass/volumetric flow rate of coolant in the PVT/PCM system [73-75,78,82,83], weather conditions [82,84] and solar irradiation [85]. In this section, the important parameters such as the type of PCM, melting temperature and thermal conductivity of PCM, type of coolant as well as position of PCM have been reviewed.

Type of PCM

The various PCMs used in the previous studies on PVT/PCM systems are reported in Table 4.3. Based on the table, most of the studies utilized organic PCMs in the PVT/PCM systems. Moreover, among the organic PCMs, paraffin waxes were more frequently examined. Based on the literature survey, only a few have investigated the effect of PCM type on the performance of PVT/PCM systems [73,74]. In a numerical study [73], the fluid temperature at the outlet of cooling system, absorber plate temperature, electrical energy efficiency, thermal energy efficiency and overall exergy efficiency of a water-based PVT/PCM system with different PCMs were compared with each other. The investigated PCMs in this study were paraffin C15, paraffin C18, paraffin C22, sodium phosphate salt and capric/palmitic acid. The results illustrated that the PVT/PCM system with paraffin C22 has a higher water temperature, and consequently, a higher thermal energy efficiency compared to those of other mentioned PVT/PCM systems. Moreover, the effect of heat capacity of PCMs on the thermal energy efficiency is more than the melting temperature and latent heat of PCMs. In addition, based on the numerical investigations, the PVT/PCM system with sodium phosphate salt has the lowest absorber plate temperature. Therefore, the electrical energy efficiency of the PVT/PCM system with sodium phosphate salt is higher than that of other studied PVT/PCM systems. In an experimental investigation [74], the effect of paraffin 42-44, Rubitherm RT22 and ceresin on the electrical performance of a water-based PVT/PCM system was evaluated. The results revealed that the PVT/PCM system with paraffin 42-44 has better performance compared to other mentioned

PVT/PCM systems. Moreover, the melting temperature and latent heat of PCMs are important parameters in order to improve the electrical performance of PV modules.

Melting Temperature of PCM

As mentioned in Table 4.3, several PCMs with different melting temperature were investigated in the previous studies on PVT/PCM systems. Selection of the melting temperature of the PCM depends on the climate and the operating conditions of the PVT system. An optimum melting temperature for a PCM is the one for which the PCM experiences both solid and liquid phases in the working time of the PVT/PCM system. The melting temperature of the PCM should be higher than the average ambient temperature during the day light in order for the PCM to be in the solid state at the ambient temperature. For example, the previous studies [75] recommended the melting temperature of 28 °C and 32 °C for the ambient temperature of 10-20 °C and 20-30 °C, respectively. Moreover, in order to cool the photovoltaic cells, the melting temperature of the PCM should be lower than the average photovoltaic cells temperature. This is an important parameter in the selection of an optimum melting temperature of the PCM. In addition, the solidification temperature of the PCM should be higher than the average ambient temperature at night. This is due to the fact that the PCM returns to the solid state the next day. A few studies have investigated the effect of melting temperature of the PCM on the performance of PVT/PCM systems [75,76]. For instance, in a numerical study [76], the effect of the melting temperature of the PCM on the photovoltaic cell temperature, fluid temperature at the outlet of cooling system, output electrical and thermal powers and electoral energy efficiency of a water-based PVT/PCM system was evaluated. In this study, the melting temperatures were 30 °C, 40 °C, 50 °C and 60 °C. The results demonstrated that the PVT/PCM system with a lower PCM melting temperature has a higher output electrical power and electrical energy efficiency.

Thermal Conductivity of PCM

The low thermal conductivity of PCMs is a major limitation of using PCMs in the PVT/PCM systems. This reduces the heat transfer rate in the PVT/PCM system. As reported in Table 4.3, the thermal conductivity of the mentioned PCMs is in the range of 0.1-0.5 W m^{-1} K^{-1}. It

Table 4.3 Different PCMs used in the previous studies on PVT/PCM systems

Reference	Material	PCM type	Melting T (°C)	Latent heat (kJ kg^{-1})	Thermal conductivity (W m^{-1} K^{-1})	Specific heat capacity (kJ kg^{-1} K^{-1})	Density (kg m^{-3})
[75,76,81]	Paraffin wax	Organic	28	210	Solid: 0.24, Liquid: 0.15	Solid: 2.9, Liquid: 2.1	Solid: 860, Liquid: 780
[78]	Paraffin wax	Organic	49	196	0.21	Solid: 2.1, Liquid: 2.1	Solid: 930, Liquid: 830
[85]	Paraffin wax	Organic	50	N/A	N/A	Solid: 1.7, Liquid: 2.2	Solid: 900, Liquid: 700
[43,80]	Paraffin wax	Organic	46-48	200-220	0.15-0.24	2.14-2.90	900
[83]	Paraffin wax RT30	Organic	28	222	0.2	Solid: 2.4, Liquid: 1.8	Solid: 870, Liquid: 760
[74]	Paraffin 42-44	Organic	42-72	130	N/A	N/A	900
[86]	n-octadecane paraffin	Organic	28	243.5	Solid: 0.358, Liquid: 0.148	Solid: 1.934, Liquid: 2.196	Solid: 865, Liquid: 770
[73]	Paraffin C15	Organic	14	205	N/A	2.52	Solid: 900, Liquid: 760
[73]	Paraffin C18	Organic	29	244	N/A	2.52	Solid: 900, Liquid: 760
[73]	Paraffin C22	Organic	44	226	N/A	Solid: 2.51, Liquid: 2.95	Solid: 818, Liquid: 760
[74]	Rubitherm RT22	Organic	20-23	200	N/A	N/A	760
[87]	Capric acid	Organic	30.1	158	0.149	Solid: 1.9, Liquid: 2.1	886
[82]	OM37	Organic	37	211	Solid: 0.50, Liquid: 0.44	Solid: 2.27, Liquid: 1.76	Solid: 960, Liquid: 862
[74]	Ceresin	Organic	61-78	160	N/A	N/A	910
[73]	Sodium phosphate salt	Inorganic	37	280	N/A	Solid: 1.94, Liquid: 1.69	1522
[84]	Salt hydrate	Inorganic	56	256	N/A	N/A	N/A
[73,88]	Capric/palmitic acid (75:25 wt%)	Eutectic	17.7-22.8	189-191	0.143	1.65	883

should be mentioned that despite the advantages of organic PCMs compared to inorganic PCMs, the thermal conductivity of inorganic PCMs (salt hydrates) is higher than that of the organic PCMs (paraffins) [12,13,74]. Based on the literature review, several attempts have been performed in order to improve the heat transfer rate in the PVT/PCM systems such as using a fin [83], employing a porous medium [73] and dispersing the nanoparticles in the PCM (referred to as Nano-PCM) [77,78]. The numerical studies [73] demonstrated that in a water-based PVT/PCM system, using copper foam as the porous medium decreases the absorber plate temperature. Consequently, the electrical energy efficiency of the PVT/PCM system is enhanced. Based on the numerical results, the electrical energy efficiency of the PVT/PCM system with copper foam is improved by 1.3% compared to the system without the porous medium.

As mentioned before, dispersing nanoparticles in the PCM is a new method in order to enhance the heat transfer rate in the PVT/PCM systems. In an experimental study [78], the effect of using the Nano-PCM on the performance of a NPVT system was evaluated. The Nano-PCM was produced by dispersing SiC nanoparticles in paraffin wax. In this study, the effect of mass fraction of nanoparticles (0.1 wt%, 0.5 wt%, 1 wt%, 2 wt%, 3 wt% and 4 wt%) on the thermophysical properties of Nano-PCM was investigated. It should be mentioned that the mass of PCM was 22 kg. The results illustrated that the thermal conductivity of the Nano-PCM increased by raising the mass fraction of nanoparticles. However, for the mass fraction higher than 0.1%, the effect of mass fraction of nanoparticles is not significant. Further research is necessary in order to investigate the effect of using different Nano-PCMs on the performance of PVT systems. Moreover, an important parameter in the Nano-PCMs is their stability for several melting/solidification cycles, as the nanoparticles may be aggregated in the melting/solidification cycles. Therefore, this subject should be investigated in future studies of PVT/Nano-PCM systems. Moreover, future research should also focus on the economic analysis of PVT/PCM and PVT/Nano-PCM systems.

Type of Coolant

The types of coolants used in the previous studies on PVT/PCM systems are mentioned in Table 4.4. According to the Table 4.4, most of the available studies used water as the coolant of the PVT/PCM system. However, only a few studies have evaluated the effect of using nanofluid in the PVT/PCM system. Therefore, more numerical and experimental investigations are required in order to study the PVT/PCM systems with different nanofluids. Moreover, the economic analysis of the NPVT/PCM systems should be considered in future investigations.

The experimental studies [87] revealed that using the PCM in a water-based PVT system improves the output electrical power of the system. Moreover, the electrical and thermal energy efficiencies of the water-based PVT/PCM system are 8.16% and 70.34%, respectively. On the other hand, the corresponding efficiencies for the PVT system are 6.98% and 58.35%, respectively. The experimental results [80] demonstrated that using water, water/EG mixture (50:50 wt%) and ethylene glycol as the coolants of a PVT/PCM system enhances the output electrical exergy of the system by 8.81%, 8.04%, and 6.13% compared to that of the PV module, respectively. Based on the experimental studies [77], the electrical energy efficiency of a NPVT/Nano-PCM system with SiC-water nanofluid (3 wt%) is 5.25% higher than that of the PV module. Moreover, the experimental investigations [43] illustrated that using ZnO-water nanofluid with 0.2 wt% as the coolant in a NPVT/PCM system enhances the output thermal exergy by 79.36% compared to that of the NPVT system. In addition, the total entropy generation of the NPVT/PCM system is 3.19% lower than the PV module. Furthermore, the output thermal exergy of the NPVT/PCM system with ZnO-water nanofluid was observed to be 23.61% higher as compared to the PVT/PCM system with pure water [79].

Table 4.4 The type of coolant used in the literature studies on PVT/PCM systems

Reference	Numerical/Experimental	Type of coolant
[81]	Numerical	Air
[73,75,76,82,86]	Numerical	Water
[74,87]	Experimental (indoor)	Water
[77-80,83-85,88]	Experimental (outdoor)	Water
[80]	Experimental (outdoor)	Ethylene glycol
[80]	Experimental (outdoor)	Water/EG mixture (50:50 wt%)
[77,78]	Experimental (outdoor)	SiC-water nanofluid
[43,79]	Experimental (outdoor)	ZnO-water nanofluid

Position of PCM

Nowadays, the phase change materials are widely used in different applications. Based on the literature survey, several research studies are available in the literature that evaluated the effect of PCMs and PVT systems in the buildings. In these studies, the PCMs are employed in the building envelop [89,90] or the building equipment such as ventilation systems [89,91]. Investigations on the PCMs in the buildings integrated with PVT systems is not of interest of this chapter, instead this section reviews different positions of the PCM in the PVT/PCM systems. In most of the previous studies on PVT/PCM systems [43,76-80,83,87,88], the collector pipes in the cooling system were surrounded by the PCM (as an example, see Figure 4.2). However, there are some research studies that presented different configurations for the PVT/PCM systems. In an experimental study [85], the efficiency of a PVT/PCM system integrated with a concentrator was evaluated. In this study, the encapsulated PCM (Paraffin wax) was immersed in a water tank. Moreover, in a numerical research [82], the effect of using an organic PCM (OM37) on the performance of a PVT system was investigated. In this PVT/PCM system, the PCM was under a fully wetted absorber channel. A new design of the PVT/PCM system was proposed in a numerical investigation [86], in which the PCM was surrounded by a channel water jacket. In a numerical study [81], the effect of the PCM position on the performance of a PVT/PCM system was investigated. In this study, two different positions were considered for the PCM: under heat transfer pipes and on top of them. Based on the results, the position of the PCM is an important parameter in a PVT/PCM system that significantly affects the performance of the system. In the PVT/PCM system, using the PCM on top of heat transfer pipes enhances the daily average electrical, thermal and overall energy efficiencies compared to using the PCM under heat transfer pipes.

4.4.2 PVT Systems with MEPCM Slurry

Literature survey reveals that there are only a few studies regarding the PVT systems with MEPCM slurry as the coolant [92-95]. In this section, the studies regarding the PVT systems with MEPCM slurry are reviewed and the investigated parameters are evaluated. The studied parameters in the literature include the thermophysical properties of MEPCM slurry [92,93], mass fraction of MEPCM [92-94], pressure drop of MEPCM slurry in the PVT system [92-94], thermal conductivity of MEPCM [94], latent heat of MEPCM [94], solar irradiation [92], mass flow rate of MEPCM slurry in the cooling system [94,95], Reynolds number [92,93] and comparison of MEPCM slurry with water [92-95].

The previous studies on PVT systems with MEPCM slurry reveal that paraffin [92,94] and octadecane [93,95] were commonly used in the MEPCMs structure. The thermophysical properties of the MEPCM slurries depend on the properties of their components (i.e., core, shell and carrier fluid) [92,93]. Only a few studies [92,93] investigated the effect of mass/volume fraction of MEPCMs on the thermophysical properties of MEPCM slurries in the PVT systems. In case, the thermal conductivity of MEPCMs components (as an example, octadecane as the core and melamine formaldehyde as the shell [93]) is lower than water (as the carrier fluid), the thermal conductivity of MEPCM slurry is decreased by increasing the mass/volume fraction of MEPCMs [92,93]. However, the previous investigations reveal that the micro-convection effect enhances the effective thermal conductivity of the MEPCM slurries [96,97]. The numerical studies [94] demonstrated that using a MEPCM with higher thermal conductivity reduces the photovoltaic cell temperature. Therefore, the electrical energy efficiency of the PVT system is improved. Based on the available studies [92,93], the viscosity of the MEPCM slurries is increased by raising the mass/volume fraction of MEPCMs. This can increase the pressure drop, and consequently, the pumping power in the PVT system [92]. The numerical investigations [94] illustrated that with a constant mass fraction of MEPCM, the electrical energy efficiency of the PVT system is increased by raising the mass flow rate of MEPCM slurry. Moreover, using the MEPCM slurry as the coolant of the PVT system enhances the daily electrical energy efficiency compared to water.

As mentioned above, only a few studies investigated the effect of MEPCM slurries on the performance of PVT systems. However, more numerical and experimental investigations are necessary in order to evaluate the electrical and thermal efficiencies of PVT systems with different MEPCM slurries from the energy and exergy viewpoints. Moreover, the effect of MEPCM size and melting temperature of PCM (as the core) on the performance of PVT systems should be investigated in future studies. In addition, the economic analysis of the PVT systems with MEPCM slurries should be considered in future investigations. It should be mentioned that several studies are available in literature that reported the effect of nano-encapsulated phase change materials (NEPCMs) in different applications [20,98,99]. However, the use of the NEPCM slurries as the coolant of PVT systems is rarely reported in the literature. Thus, future studies can be focused on the thermodynamic and economic analyses of PVT systems with NEPCM slurries. Moreover, the effect of using the nanofluids and NEPCM slurries on the performance of PVT systems can be compared with each other. As mentioned before, the dispersion of metal-oxide nanoparticles and NEPCMs in a fluid increases its thermal conductivity and latent heat storage capacity, respectively. Thus, the possibility of using a hybrid nanofluid containing the nanoparticles and NEPCMs with various mass fractions in a base fluid as the coolant of PVT systems can be investigated in future studies.

4.5 Conclusions

In this chapter, the effect of nanofluids and PCMs on the performance of PVT systems was investigated. For this purpose, the important parameters in the nanofluid-based PVT systems and PCM-based PVT systems were evaluated. In addition, in this chapter, thermodynamic (energy and exergy) analysis of the active and passive PVT systems was presented. Based on the literature review, the following subjects remain to be examined in future studies:

- More numerical and experimental studies are necessary in order to evaluate the effect of mass/volume fraction of nanoparticles as well as the particle size in the turbulent regime for various nanofluids in the NPVT systems.
- Future investigations can be focused on the study of PVT/PCM systems with different nanofluids (as the coolants).
- Only a few studies investigated the effects of PCM type on the performance of PVT/PCM systems. Thus, the thermodynamic analysis of the PVT/PCM systems with different PCM types may be the focus of future studies.
- More studies are necessary in order to investigate the effect of using different Nano-PCMs on the performance of PVT systems. Moreover, the stability of Nano-PCMs for several melting/solidification cycles should be evaluated in future studies on PVT/Nano-PCM systems.
- More numerical/experimental investigations are necessary to evaluate the performance of PVT systems with different MEPCM slurries from the energy and exergy viewpoints.

- The effect of MEPCM size and melting temperature of PCM (as the core) on the performance of PVT systems should be investigated in future studies.
- Future studies can be focused on the thermodynamic (energy and exergy) analysis of PVT systems with NEPCM slurries.
- In future studies, the effect of nanofluids and NEPCM slurries on the performance of PVT systems can be compared with each other.
- The possibility of using a hybrid nanofluid containing the nanoparticles and NEPCMs with various mass fractions in a base fluid as the coolant of PVT systems can be investigated in future investigations.
- Future investigations can be focused on the economic analysis of PVT/PCM systems, NPVT/PCM systems, PVT/Nano-PCM systems, PVT systems with MEPCM slurries and PVT systems with NEPCM slurries.

References

1. Teo, H. G., Lee, P. S., and Hawlader, M. N. A. (2012) An active cooling system for photovoltaic modules. *Applied Energy,* **90**(1), 309-315.
2. Chow, T. T. (2010) A review on photovoltaic/thermal hybrid solar technology. *Applied Energy,* **87**(2), 365-379.
3. Selimefendigil, F., Bayrak, F., and Oztop, H. F. (2018) Experimental analysis and dynamic modeling of a photovoltaic module with porous fins. *Renewable Energy,* **125**, 193-205.
4. Khanna, S., Reddy, K. S., and Mallick, T. K. (2018) Climatic behaviour of solar photovoltaic integrated with phase change material. *Energy Conversion and Management,* **166**, 590-601.
5. Dimri, N., Tiwari, A., and Tiwari, G. N. (2018) Effect of thermoelectric cooler (TEC) integrated at the base of opaque photovoltaic (PV) module to enhance an overall electrical efficiency. *Solar Energy,* **166**, 159-170.
6. Kilic, F., Menlik, T., and Sozen, A. (2018) Effect of titanium dioxide/water nanofluid use on thermal performance of the flat plate solar collector. *Solar Energy,* **164**, 101-108.
7. Yazdanifard, F., Ameri, M., and Ebrahimnia-Bajestan, E. (2017) Performance of nanofluid-based photovoltaic/thermal systems: A review. *Renewable and Sustainable Energy Reviews,* **76**, 323-352.
8. Verma, S. K., and Tiwari, A. K. (2015) Progress of nanofluid application in solar collectors: a review. *Energy Conversion and Management,* **100**, 324-346.

9. Mahian, O., Kianifar, A., Kalogirou, S. A., Pop, I., and Wongwises, S. (2013) A review of the applications of nanofluids in solar energy. *International Journal of Heat and Mass Transfer,* **57**(2), 582-594.
10. El-Samie, M. M. A., Ju, X., Xu, C., Du, X., and Zhu, Q. (2018) Numerical study of a photovoltaic/thermal hybrid system with nanofluid based spectral beam filters. *Energy Conversion and Management,* **174**, 686-704.
11. An, W., Zhang, J., Zhu, T., and Gao, N. (2016) Investigation on a spectral splitting photovoltaic/thermal hybrid system based on polypyrrole nanofluid: preliminary test. *Renewable Energy,* **86**, 633-642.
12. Islam, M. M., Pandey, A. K., Hasanuzzaman, M., and Rahim, N. A. (2016) Recent progresses and achievements in photovoltaic-phase change material technology: A review with special treatment on photovoltaic thermal-phase change material systems. *Energy Conversion and Management,* **126**, 177-204.
13. Sharma, A., Tyagi, V. V., Chen, C. R., and Buddhi, D. (2009) Review on thermal energy storage with phase change materials and applications. *Renewable and Sustainable Energy Reviews,* **13**(2), 318-345.
14. Browne, M. C., Norton, B., and McCormack, S. J. (2015) Phase change materials for photovoltaic thermal management. *Renewable and Sustainable Energy Reviews,* **47**, 762-782.
15. Waqas, A., Ji, J., Xu, L., Ali, M., Zeashan, and Alvi, J. (2018) Thermal and electrical management of photovoltaic panels using phase change materials - A review. *Renewable and Sustainable Energy Reviews,* **92**, 254-271.
16. Ma, T., Zhao, J., and Li, Z. (2018) Mathematical modelling and sensitivity analysis of solar photovoltaic panel integrated with phase change material. *Applied Energy,* **228**, 1147-1158.
17. Nada, S. A., El-Nagar, D. H., and Hussein, H. M. S. (2018) Improving the thermal regulation and efficiency enhancement of PCM-Integrated PV modules using nanoparticles. *Energy Conversion and Management,* **166**, 735-743.
18. Khanna, S., Reddy, K. S., and Mallick, T. K. (2018) Optimization of solar photovoltaic system integrated with phase change material. *Solar Energy,* **163**, 591-599.
19. Rehman, M. M. U., Qu, Z. G., Fu, R. P., and Xu, H. T. (2017). Numerical study on free-surface jet impingement cooling with nanoencapsulated phase-change material slurry and nanofluid. *International Journal of Heat and Mass Transfer,* **109**, 312-325.
20. Liu, C., Rao, Z., Zhao, J., Huo, Y., and Li, Y. (2015) Review on nanoencapsulated phase change materials: preparation, characterization and heat transfer enhancement. *Nano Energy,* **13**, 814-826.
21. Mahian, O., Kianifar, A., Kleinstreuer, C., Al-Nimr, M. A., Pop, I., Sahin, A. Z., and Wongwises, S. (2013) A review of entropy generation in

nanofluid flow. *International Journal of Heat and Mass Transfer,* **65**, 514-532.

22. Khanafer, K., and Vafai, K. (2011) A critical synthesis of thermophysical characteristics of nanofluids. *International Journal of Heat and Mass Transfer,* **54**(19-20), 4410-4428.
23. Zhou, S., and Ni, R. (2008) Measurement of the specific heat capacity of water-based Al_2O_3 nanofluid. *Applied Physics Letters,* **92**(9), 093123.
24. Bayrak, F., Abu-Hamdeh, N., Alnefaie, K. A., and Oztop, H. F. (2017) A review on exergy analysis of solar electricity production. *Renewable and Sustainable Energy Reviews,* **74**, 755-770.
25. Ebaid, M. S. Y., Ghrair, A. M., and Al-Busoul, M. (2018) Experimental investigation of cooling photovoltaic (PV) panels using (TiO_2) nanofluid in water-polyethylene glycol mixture and (Al_2O_3) nanofluid in water-cetyltrimethylammonium bromide mixture. *Energy Conversion and Management,* **155**, 324-343.
26. Joshi, A. S., Dincer, I., and Reddy, B. V. (2009) Thermodynamic assessment of photovoltaic systems. *Solar Energy,* **83**(8), 1139-1149.
27. Sardarabadi, M., Hosseinzadeh, M., Kazemian, A., and Passandideh-Fard, M. (2017) Experimental investigation of the effects of using metal-oxides/water nanofluids on a photovoltaic thermal system (PVT) from energy and exergy viewpoints. *Energy,* **138**, 682-695.
28. Vats, K., and Tiwari, G. N. (2012) Energy and exergy analysis of a building integrated semitransparent photovoltaic thermal (BISPVT) system. *Applied Energy,* **96**, 409-416.
29. Agrawal, S., and Tiwari, G. N. (2011) Energy and exergy analysis of hybrid micro-channel photovoltaic thermal module. *Solar Energy,* **85**(2), 356-370.
30. Yazdanpanahi, J., Sarhaddi, F., and Adeli, M. M. (2015) Experimental investigation of exergy efficiency of a solar photovoltaic thermal (PVT) water collector based on exergy losses. *Solar Energy,* **118**, 197-208.
31. Hasanuzzaman, M., Malek, A. B. M. A., Islam, M. M., Pandey, A. K., and Rahim, N. A. (2016) Global advancement of cooling technologies for PV systems: A review. *Solar Energy,* **137**, 25-45.
32. Chow, T. T., Pei, G., Fong, K. F., Lin, Z., Chan, A. L. S., and Ji, J. (2009) Energy and exergy analysis of photovoltaic–thermal collector with and without glass cover. *Applied Energy,* **86**(3), 310-316.
33. Hazami, M., Riahi, A., Mehdaoui, F., Nouicer, O., and Farhat, A. (2016) Energetic and exergetic performances analysis of a PV/T (photovoltaic thermal) solar system tested and simulated under to Tunisian (North Africa) climatic conditions. *Energy,* **107**, 78-94.
34. Ibrahim, A., Fudholi, A., Sopian, K., Othman, M. Y., and Ruslan, M. H. (2014) Efficiencies and improvement potential of building integrated photovoltaic thermal (BIPVT) system. *Energy Conversion and*

Management, **77**, 527-534.

35. Khanjari, Y., Pourfayaz, F., and Kasaeian, A. B. (2016) Numerical investigation on using of nanofluid in a water-cooled photovoltaic thermal system. *Energy Conversion and Management,* **122**, 263-278.
36. Khanjari, Y., Kasaeian, A. B., and Pourfayaz, F. (2017) Evaluating the environmental parameters affecting the performance of photovoltaic thermal system using nanofluid. *Applied Thermal Engineering,* **115**, 178-187.
37. Kalogirou, S. A., Karellas, S., Braimakis, K., Stanciu, C., and Badescu, V. (2016) Exergy analysis of solar thermal collectors and processes. *Progress in Energy and Combustion Science,* **56**, 106-137.
38. Evola, G., and Marletta, L. (2014) Exergy and thermoeconomic optimization of a water-cooled glazed hybrid photovoltaic/thermal (PVT) collector. *Solar Energy,* **107**, 12-25.
39. Saloux, E., Teyssedou, A., and Sorin, M. (2013) Analysis of photovoltaic (PV) and photovoltaic/thermal (PV/T) systems using the exergy method. *Energy and Buildings,* **67**, 275-285.
40. Aberoumand, S., Ghamari, S., and Shabani, B. (2018) Energy and exergy analysis of a photovoltaic thermal (PV/T) system using nanofluids: An experimental study. *Solar Energy,* **165**, 167-177.
41. Kazemian, A., Hosseinzadeh, M., Sardarabadi, M., and Passandideh-Fard, M. (2018) Effect of glass cover and working fluid on the performance of photovoltaic thermal (PVT) system: An experimental study. *Solar Energy,* **173**, 1002-1010.
42. Tiwari, A., Dubey, S., Sandhu, G. S., Sodha, M. S., and Anwar, S. I. (2009) Exergy analysis of integrated photovoltaic thermal solar water heater under constant flow rate and constant collection temperature modes. *Applied Energy,* **86**(12), 2592-2597.
43. Hosseinzadeh, M., Sardarabadi, M., and Passandideh-Fard, M. (2018) Energy and exergy analysis of nanofluid based photovoltaic thermal system integrated with phase change material. *Energy,* **147**, 636-647.
44. Al-Shamani, A. N., Sopian, K., Mat, S., Hasan, H. A., Abed, A. M., and Ruslan, M. H. (2016) Experimental studies of rectangular tube absorber photovoltaic thermal collector with various types of nanofluids under the tropical climate conditions. *Energy Conversion and Management,* **124**, 528-542.
45. Michael, J. J., and Iniyan, S. (2015) Performance analysis of a copper sheet laminated photovoltaic thermal collector using copper oxide–water nanofluid. *Solar Energy,* **119**, 439-451.
46. Rejeb, O., Sardarabadi, M., Ménézo, C., Passandideh-Fard, M., Dhaou, M. H., and Jemni, A. (2016) Numerical and model validation of uncovered nanofluid sheet and tube type photovoltaic thermal solar system. *Energy Conversion and Management,* **110**, 367-377.
47. Al-Waeli, A. H. A., Sopian, K., Chaichan, M. T., Kazem, H. A., Hasan, H.

A., and Al-Shamani, A. N. (2017) An experimental investigation of SiC nanofluid as a base-fluid for a photovoltaic thermal PV/T system. *Energy Conversion and Management,* **142**, 547-558.

48. Al-Waeli, A. H. A., Chaichan, M. T., Kazem, H. A., and Sopian, K. (2017) Comparative study to use nano-(Al_2O_3, CuO, and SiC) with water to enhance photovoltaic thermal PV/T collectors. *Energy Conversion and Management,* **148**, 963-973.
49. Yazdanifard, F., Ebrahimnia-Bajestan, E., and Ameri, M. (2017) Performance of a parabolic trough concentrating photovoltaic/thermal system: effects of flow regime, design parameters, and using nanofluids. *Energy Conversion and Management,* **148**, 1265-1277.
50. Jing, D., Hu, Y., Liu, M., Wei, J., and Guo, L. (2015) Preparation of highly dispersed nanofluid and CFD study of its utilization in a concentrating PV/T system. *Solar Energy,* **112**, 30-40.
51. Bellos, E., and Tzivanidis, C. (2017) Yearly performance of a hybrid PV operating with nanofluid. *Renewable Energy,* **113**, 867-884.
52. Maadi, S. R., Kolahan, A., Passandideh-Fard, M., Sardarabadi, M., and Moloudi, R. (2017) Characterization of PVT systems equipped with nanofluids-based collector from entropy generation. *Energy Conversion and Management,* **150**, 515-531.
53. Al-Waeli, A. H. **A.**, Chaichan, M. T., Kazem, H. A., Sopian, K., Ibrahim, A., Mat, S., and Ruslan, M. H. (2018) Comparison study of indoor/outdoor experiments of a photovoltaic thermal PV/T system containing SiC nanofluid as a coolant. *Energy,* **151**, 33-44.
54. Al-Shamani, A. N., Sopian, K., Mat, S., and Abed, A. M. (2017) Performance enhancement of photovoltaic grid-connected system using PVT panels with nanofluid. *Solar Energy,* **150**, 38-48.
55. Ghadiri, M., Sardarabadi, M., Pasandideh-Fard, M., and Moghadam, A. J. (2015) Experimental investigation of a PVT system performance using nano ferrofluids. *Energy Conversion and Management,* **103**, 468-476.
56. Karami, N., and Rahimi, M. (2014) Heat transfer enhancement in a PV cell using Boehmite nanofluid. *Energy Conversion and Management,* **86**, 275-285.
57. Heidari, N., Rahimi, M., and Azimi, N. (2018) Experimental investigation on using ferrofluid and rotating magnetic field (RMF) for cooling enhancement in a photovoltaic cell. *International Communications in Heat and Mass Transfer,* **94**, 32-38.
58. Moradgholi, M., Nowee, S. M., and Farzaneh, A. (2018) Experimental study of using Al_2O_3/methanol nanofluid in a two phase closed thermosyphon (TPCT) array as a novel photovoltaic/thermal system. *Solar Energy,* **164**, 243-250.
59. Rostami, Z., Rahimi, M., and Azimi, N. (2018) Using high-frequency ultrasound waves and nanofluid for increasing the efficiency and cooling performance of a PV module. *Energy Conversion and Manag-*

ement, **160**, 141-149.

60. Fayaz, H., Nasrin, R., Rahim, N. A., and Hasanuzzaman, M. (2018) Energy and exergy analysis of the PVT system: Effect of nanofluid flow rate. *Solar Energy*, **169**, 217-230.
61. Youssef, W. B., Maatallah, T., Menezo, C., and Nasrallah, S. B. (2018) Modeling and optimization of a solar system based on concentrating photovoltaic/thermal collector. *Solar Energy*, **170**, 301-313.
62. Bianco, V., Scarpa, F., and Tagliafico, L. A. (2018) Numerical analysis of the Al_2O_3-water nanofluid forced laminar convection in an asymmetric heated channel for application in flat plate PV/T collector. *Renewable Energy*, **116**, 9-21.
63. Nasrin, R., Rahim, N. A., Fayaz, H., and Hasanuzzaman, M. (2018) Water/MWCNT nanofluid based cooling system of PVT: Experimental and numerical research. *Renewable Energy*, **121**, 286-300.
64. Sardarabadi, M., Passandideh-Fard, M., and Heris, S. Z. (2014) Experimental investigation of the effects of silica/water nanofluid on PV/T (photovoltaic thermal units). *Energy*, **66**, 264-272.
65. Hosseinzadeh, M., Salari, A., Sardarabadi, M., and Passandideh-Fard, M. (2018) Optimization and parametric analysis of a nanofluid based photovoltaic thermal system: 3D numerical model with experimental validation. *Energy Conversion and Management*, **160**, 93-108.
66. Radwan, A., and Ahmed, M. (2018) Thermal management of concentrator photovoltaic systems using microchannel heat sink with nanofluids. *Solar Energy*, **171**, 229-246.
67. Lekbir, A., Hassani, S., Ghani, M. R. A., Gan, C. K., Mekhilef, S., and Saidur, R. (2018) Improved energy conversion performance of a novel design of concentrated photovoltaic system combined with thermoelectric generator with advance cooling system. *Energy Conversion and Management*, **177**, 19-29.
68. Al-Waeli, A. H. A., Chaichan, M. T., Kazem, H. A., Sopian, K., and Safaei, J. (2018) Numerical study on the effect of operating nanofluids of photovoltaic thermal system (PV/T) on the convective heat transfer. *Case Studies in Thermal Engineering*, **12**, 405-413.
69. Sardarabadi, M., and Passandideh-Fard, M. (2016) Experimental and numerical study of metal-oxides/water nanofluids as coolant in photovoltaic thermal systems (PVT). *Solar Energy Materials and Solar Cells*, **157**, 533-542.
70. Hasan, H. A., Sopian, K., Jaaz, A. H., and Al-Shamani, A. N. (2017) Experimental investigation of jet array nanofluids impingement in photovoltaic/thermal collector. *Solar Energy*, **144**, 321-334.
71. Al-Shamani, A. N., Alghoul, M. A., Elbreki, A. M., Ammar, A. A., Abed, A. M., and Sopian, K. (2018) Mathematical and experimental evaluation of thermal and electrical efficiency of PV/T collector using different water based nano-fluids. *Energy*, **145**, 770-792.

72. Xu, Z., and Kleinstreuer, C. (2014) Concentration photovoltaic–thermal energy co-generation system using nanofluids for cooling and heating. *Energy Conversion and Management,* **87**, 504-512.
73. Mousavi, S., Kasaeian, A., Shafii, M. B., and Jahangir, M. H. (2018) Numerical investigation of the effects of a copper foam filled with phase change materials in a water-cooled photovoltaic/thermal system. *Energy Conversion and Management,* **163**, 187-195.
74. Klugmann-Radziemska, E., and Wcisło-Kucharek, P. (2017) Photovoltaic module temperature stabilization with the use of phase change materials. *Solar Energy,* **150**, 538-545.
75. Malvi, C. S., Dixon-Hardy, D. W., and Crook, R. (2011) Energy balance model of combined photovoltaic solar-thermal system incorporating phase change material. *Solar Energy,* **85**(7), 1440-1446.
76. Su, D., Jia, Y., Lin, Y., and Fang, G. (2017) Maximizing the energy output of a photovoltaic–thermal solar collector incorporating phase change materials. *Energy and Buildings,* **153**, 382-391.
77. Al-Waeli, A. H. A., Sopian, K., Kazem, H. A., Yousif, J. H., Chaichan, M. T., Ibrahim, A., Mat, S., Ruslan, M. H. (2018) Comparison of prediction methods of PV/T nanofluid and nano-PCM system using a measured dataset and artificial neural network. *Solar Energy,* **162**, 378-396.
78. Al-Waeli, A. H. A., Sopian, K., Chaichan, M. T., Kazem, H. A., Ibrahim, A., Mat, S., and Ruslan, M. H. (2017) Evaluation of the nanofluid and nano-PCM based photovoltaic thermal (PVT) system: An experimental study. *Energy Conversion and Management,* **151**, 693-708.
79. Sardarabadi, M., Passandideh-Fard, M., Maghrebi, M., and Ghazikhani, M. (2017) Experimental study of using both ZnO/water nanofluid and phase change material (PCM) in photovoltaic thermal systems. *Solar Energy Materials and Solar Cells,* **161**, 62-69.
80. Kazemian, A., Hosseinzadeh, M., Sardarabadi, M., and Passandideh-Fard, M. (2018) Experimental study of using both ethylene glycol and phase change material as coolant in photovoltaic thermal systems (PVT) from energy, exergy and entropy generation viewpoints. *Energy,* **162**, 210-223.
81. Su, D., Jia, Y., Alva, G., Liu, L., and Fang, G. (2017) Comparative analyses on dynamic performances of photovoltaic–thermal solar collectors integrated with phase change materials. *Energy Conversion and Management,* **131**, 79-89.
82. Gaur, A., Menezo, C., and Giroux-Julien, S. (2017) Numerical studies on thermal and electrical performance of a fully wetted absorber PVT collector with PCM as a storage medium. *Renewable Energy,* **109**, 168-187.
83. Preet, S., Bhushan, B., and Mahajan, T. (2017) Experimental investigation of water based photovoltaic/thermal (PV/T) system with and without phase change material (PCM). *Solar Energy,* **155**, 1104-

1120.

84. Al Imam, M. **F.**, Beg, R. A., Rahman, M. S., and Khan, M. Z. H. (2016) Performance of PVT solar collector with compound parabolic concentrator and phase change materials. *Energy and Buildings,* **113**, 139-144.
85. Su, Y., Zhang, Y., and Shu, L. (2018) Experimental study of using phase change material cooling in a solar tracking concentrated photovoltaic-thermal system. *Solar Energy,* **159**, 777-785.
86. Emam, M., and Ahmed, M. (2018) Performance analysis of a new concentrator photovoltaic system integrated with phase change material and water jacket. *Solar Energy,* **173**, 1158-1172.
87. Yang, X., Sun, L., Yuan, Y., Zhao, X., and Cao, X. (2018) Experimental investigation on performance comparison of PV/T-PCM system and PV/T system. *Renewable Energy,* **119**, 152-159.
88. Browne, M. C., Norton, B., and McCormack, S. J. (2016) Heat retention of a photovoltaic/thermal collector with PCM. *Solar Energy,* **133**, 533-548.
89. Mengjie, S., Fuxin, N., Ning, M., Yanxin, H., and Shiming, D. (2017) Review on building energy performance improvement using phase change materials. *Energy and Buildings,* **158**, 776–793.
90. Lin, W., and Ma, Z. (2016) Using Taguchi-Fibonacci search method to optimize phase change materials enhanced buildings with integrated solar photovoltaic thermal collectors. *Energy,* **106**, 23-37.
91. Lin, W., Ma, Z., Sohel, M. I., and Cooper, P. (2014) Development and evaluation of a ceiling ventilation system enhanced by solar photovoltaic thermal collectors and phase change materials. *Energy Conversion and Management,* **88**, 218-230.
92. Qiu, Z., Ma, X., Zhao, X., Li, P., and Ali, S. (2016) Experimental investigation of the energy performance of a novel Micro-encapsulated Phase Change Material (MPCM) slurry based PV/T system. *Applied Energy,* **165**, 260-271.
93. Qiu, Z., Zhao, X., Li, P., Zhang, X., Ali, S., and Tan, J. (2015) Theoretical investigation of the energy performance of a novel MPCM (microencapsulated phase change material) slurry based PV/T module. *Energy,* **87**, 686-698.
94. Liu, L., Jia, Y., Lin, Y., Alva, G., and Fang, G. (2017) Performance evaluation of a novel solar photovoltaic–thermal collector with dual channel using microencapsulated phase change slurry as cooling fluid. *Energy Conversion and Management,* **145**, 30-40.
95. Liu, L., Jia, Y., Lin, Y., Alva, G., and Fang, G. (2017) Numerical study of a novel miniature compound parabolic concentrating photovoltaic/thermal collector with microencapsulated phase change slurry. *Energy Conversion and Management,* **153**, 106-114.
96. Qiu, Z., Ma, X., Li, P., Zhao, X., and Wright, A. (2017) Micro-encapsulated phase change material (MPCM) slurries: Characterization and

building applications. *Renewable and Sustainable Energy Reviews,* **77**, 246-262.

97. Zhang, C. Y., Wang, T., Chen, D. H., Hong, F. J., and Cheng, P. (2016) Confined jet array impingement cooling with spent flow distraction using NEPCM slurry. *International Communications in Heat and Mass Transfer,* **77**, 140-147.
98. Wu, W., Bostanci, H., Chow, L. C., Ding, S. J., Hong, Y., Su, M., Kizito, J. P., Gschwender, L., Snyder, C. E. (2011) Jet impingement and spray cooling using slurry of nanoencapsulated phase change materials. *International Journal of Heat and Mass Transfer,* **54**(13-14), 2715-2723.
99. Seyf, H. R., Zhou, Z., Ma, H. B., and Zhang, Y. (2013) Three dimensional numerical study of heat-transfer enhancement by nano-encapsulated phase change material slurry in microtube heat sinks with tangential impingement. *International Journal of Heat and Mass Transfer,* **56**(1-2), 561-573.

5

Inorganic Phase Change Materials: A Review

Gigi George

Department of Chemistry, CMS College Kottayam, Mahatma Gandhi University, Kerala 686001, India

gigi@cmscollege.ac.in

5.1 Introduction

Devices with increasing power, smaller footprints and greater mobility are demanding cost effective thermal management solutions and many options are available today. Amongst them, one technology is attracting a variety of military and industrial applications. Use of temporary thermal storage materials or phase change materials, also known as PCMs, represents this effective technology. PCMs can assimilate thermal energy and store it amid solid to liquid stage changes. This enables the temperature to be maintained close to the melting point of the PCM. Melting and freezing are very common material properties but the useful capability of the PCMs is that the materials can be used to absorb or release energy in the form of heat at the same time which is generally a function of the kind of material it is made from, as given in Figure 5.1. During melting, the material absorbs energy and while freezing it releases energy. In the latter case, it is the latent heat which is released while absorption of energy from the surroundings takes place in the former case. In both cases, the mechanism happens at a constant temperature and the fluctuation from this will make the PCMs recharge. This particular property is very useful for such materials' applicability in many routine as well as advanced applications. However, some of the common PCMs such as water/ice cannot be used over a wide temperature range, which is a limitation. Expanding the usefulness to a wider temperature range (such as from -50 °C to more than 160 °C) would be useful for many applications. In addition, the heat storage can be achieved over a finer temperature window [1-10].

Phase Change Materials, edited by Vikas Mittal

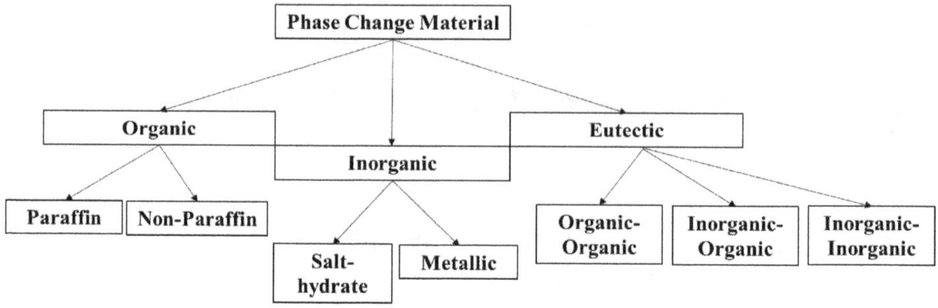

Figure 5.1 Classification of PCMs.

5.2 Properties and Performance

Based on the way in which the PCMs function and respond to the physical changes, these are classified as solid-solid, solid-liquid and liquid-gas. Phase changing materials are becoming increasingly popular in an era of energy sources getting depleted, along with simultaneously rising concerns over environmental pollution. Clean energy, which can be stored and reused, has become an interesting and promising alternative around the world. The energy storage using materials can be achieved using different methods such as thermal, chemical, electrical and mechanical. The latent heat storage has potential in many energy storage applications depending on the material property when exposed to heat and its subsequent phase transformation.

Due to their efficient performance, inorganic phase changing materials are considered in the current review. Figure 5.2 and 5.3 present the melting point and heat of fusion values of different inorganic PCMs. Figure 5.4 also depicts their thermal conductivity.

The absorption or release of energy can be mathematically represented as

$$Q = \int_{-Ti}^{Tm} mCpdT + ma_m\Delta h_m + \int_{Tm}^{Tf} mCpdT \tag{1}$$

where m is the mass; Cp is the specific heat capacity; T the temperature; a_m the melted fraction; Δh_m the enthalpy of fusion. There are various ideal physical and thermal characteristics which are considered favorable which include degradation behaviors, thermal conductivity, super-cooling behaviors, density, heat capacity, changes in

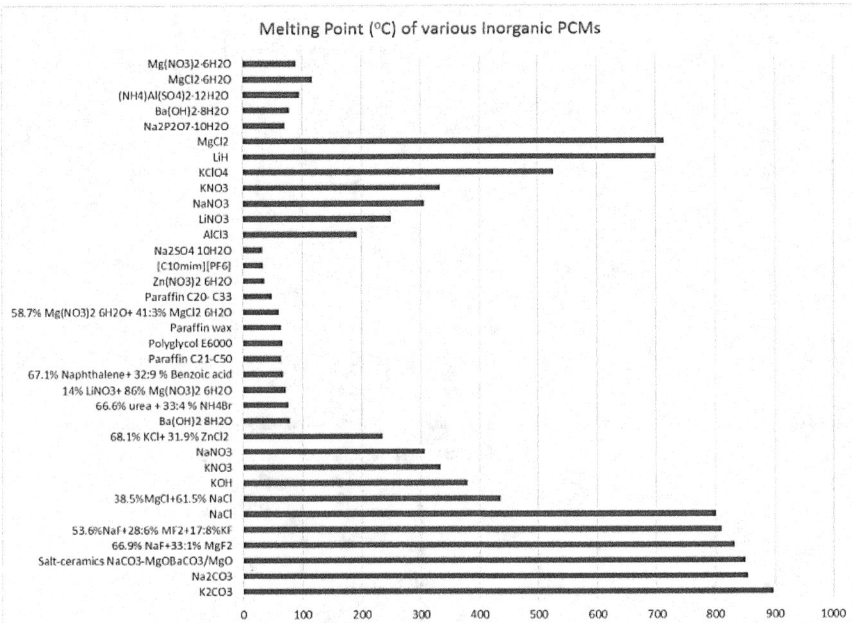

Figure 5.2 Melting point of various inorganic PCMs.

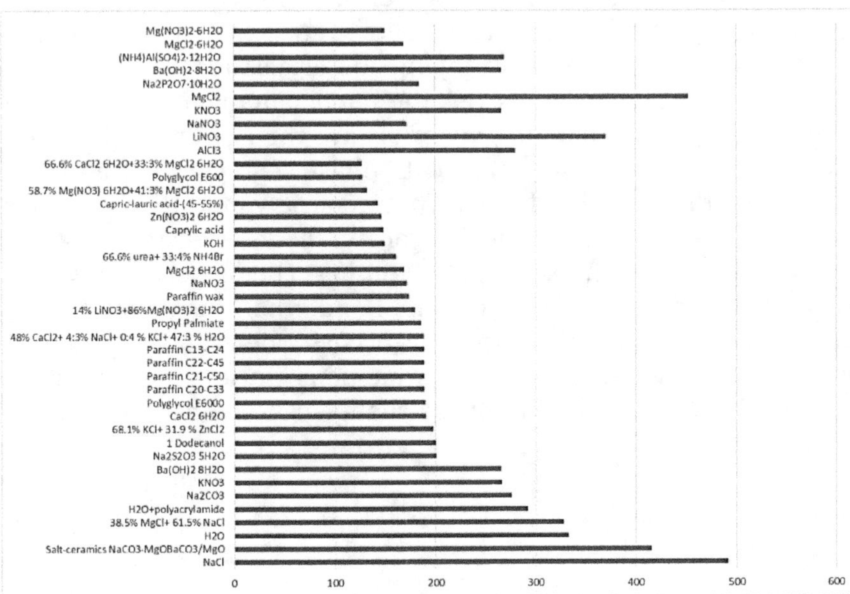

Figure 5.3 Heat of fusion of various inorganic PCMs.

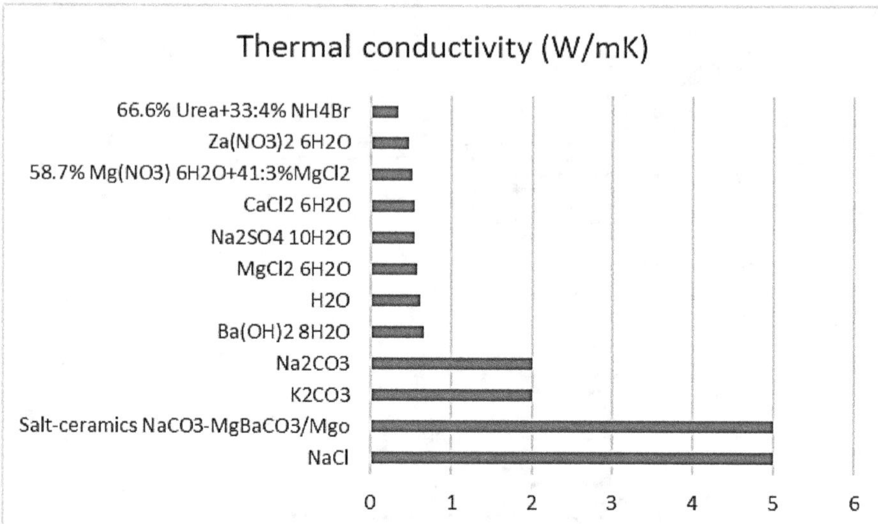

Figure 5.4 Thermal conductivity of various inorganic PCMs.

volume, chemical stability, melting temperature, crystallization temperature, safety aspects, toxicity, availability, cost effectiveness, etc.

The inorganic phase changing materials discussed in this review include salts, salt hydrates, metals alloys, etc. They are considered better than organic phase changing materials due to their high phase change enthalpy. However, the major anomalies of the inorganic class of phase changing materials are their poor stability and corrosion resistance. Also, issues with phase segregation are another challenge these materials face. They generally hold the molecular formula $X_aY_b.nH_2O$; where the X_aY_b are carbonates of metal which are formed through the ion-dipole bonding. The loosely held water molecules are susceptible to lose the bond when the phase transition of the materials takes place. This occurs when the materials absorb the latent heat. The opposite happens during the hydration process. The similar trend in phase change can be observed during the melting-freezing process.

$$XY.nH_2O \rightarrow XY.mH_2O+(n-m)\ H_2O \quad (2)$$
$$XY.nH_2O \rightarrow XY+ nH_2O \quad (3)$$

PCM materials offer many advantages but have to overcome many issues when it comes to factors concerned with practical applications. Some of the vital challenges are listed below, along with some of the strategies to address them:

a. Volume change during phase transition which in some cases was reported to be more than 10%, which has been studied over its eutectic composition for improvement [11,12].
b. Shortened thermal conductivity, which has been addressed by modulating the heat transfer area and through proper material selection.
c. Effect of super-cooling were addressed with use of external heat control devices, chemicals, etc. [1,13]
d. Corrosion issues
e. Varying melt temperature [14]
f. Cost
g. Environmental concerns [2]

Latent heat of phase change materials can be modified when they are in nanoscale. Zhang *et al.* [8] reported an increased phase change latent heat of stearic acid via nanocapsule interface confinement and the schematic of the approach is presented in Figure 5.5. The heat storage capacity of the obtained SA@SiO_2 nanocapsules (NCs) could be increased up to 374.2 kJ/k.

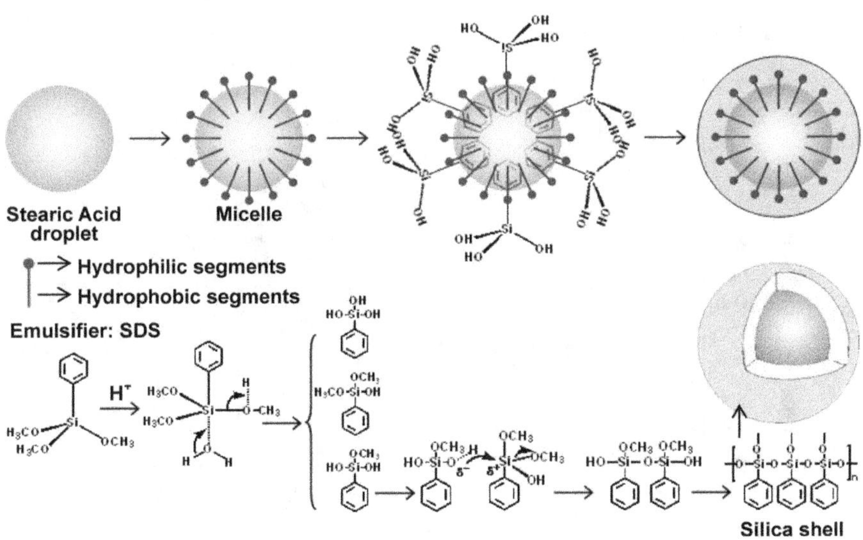

Figure 5.5 Schematic formation of the SA@SiO_2 NCs. Reproduced from Reference 8 with permission from American Chemical Society.

The phase transformation of eutectic and non-eutectic materials during the melting and freezing process need to be under control so that the PCMs can suppress their natural challenges. Recent experimental research shows that inorganic non-eutectic salt hydrates have also high potential towards overcoming the general drawbacks. Low heat of fusion and high material weight are the major limitations of this class of PCM materials. However, high thermal conductivity and flexibility to enhance its thermal conductivity based on material selection are key advantages.

Earlier studies were mostly limited to organic PCMs. The common test technique is differential scanning calorimetry which demands a very small but pure and homogenous test sample. The inorganic PCMs can adapt to other testing method, such as T-history method, which can provide better insights into thermal diffusivity and enthalpy calculations, along with the freedom of using various sample sizes [15,16]. For instance, the thermal conductivity of calcium chloride hexahydrate with γ-Al_2O_3 nanocomposite PCM was studied and was found to have an acceptable low-temperature solar thermal energy storage capability, as shown in figure 5.6 [15]. Overall, lack of standard testing method for PCM materials, regardless of the type of material, is a major challenge in evaluating the material performance.

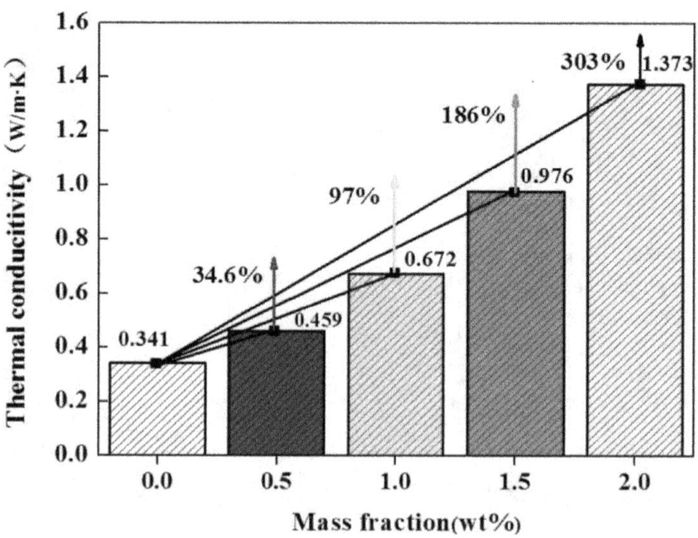

Figure 5.6 Thermal conductivity of nanocomposite PCMs. Reproduced from Reference 15 with permission from American Chemical Society.

5.3 Service Life Enhancement of PCMs

The system design, material corrosion and the physical hindrances that happen due to the repeated cycling of phase transformations can affect the life stability.

Corrosion: The corrosion of the PCM ampules is considered to be a major issue as this affects the life time. Enhancements can be achieved by systematic research on corrosion mechanisms such as oxidation, pitting, de-alloying, etc. Steel, aluminum and zinc are the metals which are generally used with conventional inorganic salt hydrate PCMs such as $CaCl_2$, Na_2SO_4 etc.

Thermal Cycling: Like other physical characteristics such as corrosion and stability, thermal properties have to be tested for a PCM material for a life time assessment. The life time of a PCM is dependent on its application. For instance, a life time of around 20-30 years is expected for advanced application such as aerospace. The thermal properties such as melt and fusion temperatures do not show any significant relation but the thermal recycling is vital to be tested as it contributes significantly to the life span of PCMs. The melt and fusion temperatures have to stay unaltered after repeated cycling and recycling process. The material selection for manufacturing and life time calculations has also direct implication on the cost effectiveness of the PCM.

System Design: One of the major design parameters that helps in enhancing the life span of a PCM is the encapsulation method by which the material is integrated into the structure. This will protect the PCM from external effects due to variations in outer environment, temperature and corrosion. Also, the internal melt-fuse phase change may lead to leakage from the shell which could be unfavorable for extended life span and efficiency of PCMs. For this purpose, a thermoregulated composite fiber with core-sheath structure containing acrylic-based copolymer microencapsulated phase-change materials was fabricated by Li *et al.* [17]. Scanning electron micrographs of fibers containing various PCM fractions are shown in Figure 5.7 [17]. The integration method of PCM material to the structure is important and there are physical and chemical methods by which this can be achieved. Physical method includes spray drying, centrifugal and fluidized bed, whereas the chemical method includes *in-situ* polymerization, interfacial polymerization and complex coacervation.

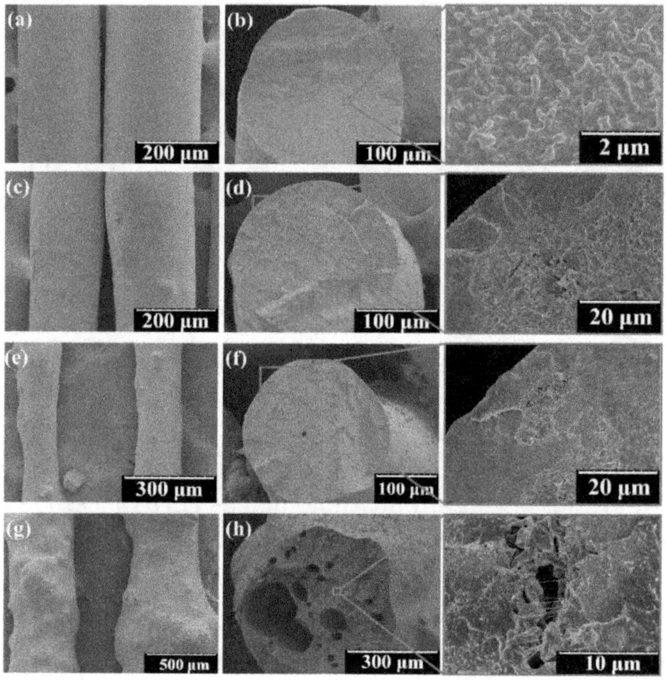

Figure 5.7 Scanning electron micrographs of fibers with various PCM amounts: (a and b) 0 wt%; (c and d) 5 wt%; (e and f) 10 wt%; (g and h) 15 wt%. Reproduced from Reference 17 with permission from American Chemical Society.

5.4 Thermal Conductivity Enhancement of PCMs

The thermal conductivity of inorganic PCMs is generally higher than organic PCMs, however, it is still low, which is one of the principle downsides for their application in numerous thermal frameworks. The most well-known strategies to improve thermal conductivity include the utilization of a broadened metal network structures, utilization of the impregnation of permeable materials and utilization of heat funnels.

Surface/metal grid PCMs help incrementing the capacity framework by incorporating different shapes with the PCMs compartments to build an exchange region. This can be achieved by using materials such as graphite, steel, aluminum and copper. The material property factors that are needed to be considered include conductivity, thickness, cost and erosion capability. The material geometry also plays a vital role.

Sodium acetate trihydrate can be used as a PCM material which can be made into geometries with and without fin network, where fins enable enhanced thermal conductivity [6]. Sodium nitrate has also been used in combination with graphite and aluminum. A laboratory test revealed more than 150 recycles for 400 hours at temperatures above 250 °C [3]. In combination with graphite, the corrosion and disintegration issues were minimal. In a similar study, $KNO_3/NaNO_3$ was used as a PCM material in combination with graphite fins [18]. Another inorganic molten material mixture including a majority constituent as LiF and minimal quality of CaF_2 was used as a PCM material and the performance of storage efficiency was enhanced by geometrical changes such as helically-finned tube [19]. A mathematical approach to energy storage mechanism and its efficiency was developed by Tiari *et al.* [20]. According to the study by Mesalhy *et al.* [21], the improvement in thermal storage capabilities can be achieved by incorporation of porous materials by systematic heat transfer and melt-fuse rates. $NaNO_3$ using graphite on porous metal sheets as PCM was studied in a similar work by Wu and Zhao [22] and the study revealed a 2.5 times performance increment as compared to the case without porous networks.

Critical research demonstrated that PCM-concrete has better inert warmth stockpiling and warm execution. Tao *et al.* [23] described the use of calcium chloride hexahydrate along with additives as PCMs in joining concrete. In another study, unique advantages of using PCM materials as automotive component contributors were discussed. PCM-graphite mix was used as a component for enhancing the life of automotive battery [10]. A numerical approach to describe the phase change of an inorganic PCM containing carbon fibers was also illustrated by Frusteri [24]. In other recent studies, Ushak *et al.* [25,26] reported the reparation and characterization of non-metallic inorganic PCMs.

5.5 Conclusions

To conclude, inorganic PCMs have appealing attributes. They portray a wide liquefying temperature run, along with thermal protection and high idle warmth rate. However, the deficient long haul steadiness of these materials is still an issue, which has restricted their utilization as warmth stockpiling frameworks. This has been accounted within the consideration of two variables such as the corrosion and thermal cycling, and forms the direction of ongoing research.

Thus, inorganic PCMs have high potential towards commercialization but require for fundamental insights into material behavior and cost effectiveness.

References

1. Telkes, M., (1980) Thermal storage in salt-hydrates. *Solar Energy Materials*, **2**(4), 381-393.
2. Abe, Y., Takahashi, Y., Kanari, K., Tanaka, K., Sakamoto, R., and Kamimoto, M. (1989) Molten Salt Latent Thermal Storage using NaOH-based Eutectics. *Proceedings of the 23rd Intersociety Energy Conversion Engineering Conference*, USA, pp. 159-164.
3. Steinmann, W.-D., Laing, D., and Tamme, R. (2009) Development of PCM storage for process heat and power generation. *Journal of Solar Energy Engineering*, **131**(4), 041009.
4. Castellon, C., Martorell, I., Cabeza, L. F., Fernandez, A. I., and Manich, A. (2011) Compatibility of plastic with phase change materials (PCM). *International Journal of Energy Research*, **35**(9), 765-771.
5. Sivasamy, P., Devaraju, A., and Harikrishnan, S. (2018) Review on heat transfer enhancement of phase change materials (PCMs). *Materials Today: Proceedings*, **5**(6), 14423-14431.
6. Choi, J. C., Kim, S. D., and Han, G. Y. (1996) Heat transfer characteristics in low-temperature latent heat storage systems using salt-hydrates at heat recovery stage. *Solar Energy Materials and Solar Cells*, **40**(1), 71-87.
7. Zalba, B., Marin, J. M., Cabeza, L. F., and Mehling, H. (2003) Review on thermal energy storage with phase change: materials, heat transfer analysis and applications. *Applied Thermal Engineering*, **23**(3), 251-283.
8. Zhang, S., Wang, S., Zhang, J., Jiang, Y., Ji, Q., Zhang, Z., and Wang, Z. (2013) Increasing phase change latent heat of stearic acid via nanocapsule interface confinement. *The Journal of Physical Chemistry C*, **117**(44), 23412-23417.
9. Luan, Y., Yang, M., Ma, Q., Qi, Y., Gao, H., Wu, Z., and Wang, G. (2016) Introduction of an organic acid phase changing material into metal–organic frameworks and the study of its thermal properties. *Journal of Materials Chemistry A*, **4**(20), 7641-7649.
10. Jaguemont, J., Omar, N., Van den Bossche, P., and Van Mierlo, J.

(2017) Phase-change materials (PCM) for automotive applications: a review. *Applied Thermal Engineering*, **132**, 308-320.

11. Kenisarin, M. M. (2010) High-temperature phase change materials for thermal energy storage. *Renewable and Sustainable Energy Reviews*, **14**(3), 955-970.
12. Kong, Q., Ma, J., and Che, C. (2009) Theoretical and experimental study of volumetric change rate during phase change process. *International Journal of Energy Research*, **33**(5), 513-525.
13. Khadiran, T., Hussein, M. Z., Zainal, Z., and Rusli, R. (2016) Advanced energy storage materials for building applications and their thermal performance characterization: A review. *Renewable and Sustainable Energy Reviews*, **57**, 916-928.
14. Heidenreich, G., and Parekh. M. (1986) Thermal Energy Storage for Organic Rankine Cycle Solar Dynamic Space Power Systems. *Proceedings of the Twenty-first Intersociety Energy Conversion Engineering Conference*, USA, pp. 791-797.
15. Li, X., Zhou, Y., Nian, H., Zhang, X., Dong, O., Ren, X., Zeng, J., Hai, C., and Shen, Y. (2017) Advanced Nanocomposite Phase Change Material Based on Calcium Chloride Hexahydrate with Aluminum Oxide Nanoparticles for Thermal Energy Storage. *Energy and Fuels*, **31**(6), 6560-6567.
16. Fu, X., Liu, Y., Jiang, X., Wang, Q., Luo, Y., and Lyu, Y. (2018) Form-stable phase change nanocomposites for thermal energy storage based on hypercrosslinked polymer nanospheres. *Thermochimica Acta*, **665**, 111-118.
17. Li, W., Ma, Y.-j., Tang, X.-f., Jiang, N., Zhang, R., Han, N., and Zhang, X.-x. (2014) Composition and characterization of thermoregulated fiber containing acrylic-based copolymer microencapsulated phase-change materials (MicroPCMs). *Industrial & Engineering Chemistry Research*, **53**(13), 5413-5420.
18. Bayon, R., Rojas, E., Valenzuela, L., Zarza, E., and León, J. (2010) Analysis of the experimental behaviour of a 100 kWth latent heat storage system for direct steam generation in solar thermal power plants. *Applied thermal engineering*, **30**(17-18), 2643-2651.
19. Tao, Y., He, Y. and Qu, Z. (2012) Numerical study on performance of molten salt phase change thermal energy storage system with enhanced tubes. *Solar energy*, **86**(5), 1155-1163.
20. Tiari, S., Qiu, S., and Mahdavi, M. (2015) Numerical study of finned heat pipe-assisted thermal energy storage system with high temperature phase change material. *Energy Conversion and Management*, **89**, 833-842.
21. Mesalhy, O., Lafdi, K., Elgafy, A., and Bowman, K. (2005) Numerical study for enhancing the thermal conductivity of phase change material (PCM) storage using high thermal conductivity porous matrix. *Energy Conversion and Management*, **46**(6), 847-867.

22. Wu, Z., and Zhao, C. (2011) Experimental investigations of porous materials in high temperature thermal energy storage systems. *Solar energy*, **85**(7), 1371-1380.
23. Tao, N., and Huang. H. (2018) Application of Phase Change Material (PCM) in Concrete for Thermal Energy Storage. *International Congress on Polymers in Concrete*, pp. 187-193.
24. Frusteri, F., Leonardi, V., and Maggio, G. (2006) Numerical approach to describe the phase change of an inorganic PCM containing carbon fibres. *Applied Thermal Engineering*, **26**(16), 1883-1892.
25. Ushak, S., Gutierrez, A., Galleguillos, H., Fernandez, A. G., Cabeza, L. F., and Grageda, M. (2015) Thermophysical characterization of a by-product from the non-metallic industry as inorganic PCM. *Solar Energy Materials and Solar Cells*, **132**, 385-391.
26. Ushak, S., Cruz, M. J. Cabeza, L. F. and Grageda, M. (2016) Preparation and characterization of inorganic PCM microcapsules by fluidized bed method. *Materials*, **9**(1), 24.

6

Attapulgite Based Composite Phase Change Material for Solar Thermal Energy Storage

Ahmet Sari,[1,2,*] Alper Biçer,[3] Mohammed Ouikhalfan[1,4] and Gökhan Hekimoğlu[1]

[1]*Karadeniz Technical University, Department of Metallurgical and Material Engineering, 61080, Trabzon, Turkey*
[2]*Centers of Research Excellence, Renewable Energy Research Institute, KFUPM, 31261, Dhahran, KSA*
[3]*Gaziosmanpaşa University, Department of Chemistry, 60240 Tokat, Turkey*
[4]*Cadi Ayyad University, Department of Applied Physics, Marrakesh, Morocco*

**Corresponding author*: ahmet.sari@ktu.edu.tr

6.1 Introduction

Daily human activities need energy as a main source of power, and its conservation through efficient utilization is vital for a sustainable economy of any country. Moreover, fossil fuels are limited and are depleting at a fast rate, and excessive usage causing environmental pollution is an international primary concern. Cutting the dependence on fossil fuels has been undertaken by the majority of world countries, using technologies involving alternate energy resources. Most of these countries rely on economic policies incenting for the extensive utilization of alternate or renewable energy [1].

The member countries of the Organization for Economic Cooperation and Development have made efforts to switch to renewable energy based technologies in order to diminish the gap between power demand and supply. The global energy demand is increasing by 1.4% every year and the buildings sector, both residential and commercial, constitute 40% of global energy consumption [2]. In terms of abundance, solar energy has been the primary energy source as it is available to the entire globe over the year and the energy received from the sun by earth is around 10,000 times of world energy demand [3]. A phase change material (PCM) is a ther-

Phase Change Materials, edited by Vikas Mittal

mal storage medium that enables the use of solar energy and decreases active energy consumption in buildings for heating, ventilation and air conditioning (HVAC). This can be reached through the application of appropriate thermal energy storage (TES) technique on building compartments such as walls, floors and ceilings as well as on building envelopes.

6.1.1 Thermal Energy Storage

The implementation TES can be achieved by using a single phase medium as sensible heat (liquid or solid), a phase changeable medium as latent heat or a reversible chemical reaction as chemical energy. The advantage of sensible heat storage is the relative ease of application by raising or decreasing the substance's temperature. Nevertheless, its heat storage capacity is generally low and the used material requires large weight or volume, which is limiting for its utilization, as for chemical energy storage, the technical and economic difficulties narrow its application. Currently, latent heat storage using PCMs has been the subject of numerous investigations and potential applications of TES.

6.1.2 Phase Change Materials

Latent heat storage with PCMs can be achieved through solid to liquid, solid to gas, liquid to gas and solid to solid. Solid to liquid transition gained popularity for application in various areas because of the issues associated with other phase change processes. Liquid to gas transition requires large volumes and high pressures making it impractical. Solid to solid transition provides low latent heat of fusion even though it is compact in nature. Solid to gas transition also has the same disadvantage as the liquid to gas transition [4].

PCMs with solid-liquid phase change are widely used due to many advantages, like minor volume change, simple handling, compactness, etc. There is a variety of substances that can be used as solid-liquid PCMs. PCM changes its phase at certain range of temperature and during phase change it absorbs or releases heat, and when incorporated into building envelope, it serves as passive thermal regulator of indoor temperature.

The PCMs can be organic, inorganic or eutectic materials. Inorganic PCMs are characterized with high latent heat per unit mass and relatively high thermal conductivity. Typical inorganic PCMs include water, salt hydrate, aqueous solution, molten salt, etc. The main drawbacks of inorganic PCM are phase segregation and incongruent phase change. Usually, inorganic PCMs are cheaper than organic PCMs. Organic PCMs include paraffins and non-paraffins. The latent heat of fusion and melting point of paraffins are dependent on the chain length. The melting point temperature of these PCMs ranges from 12-71 °C. Furthermore, these are corrosion-resistant and are free from sub-cooling. These are inexpensive, reliable and chemically stable. Major disadvantages of paraffins include low thermal conductivity, leaking from matrix, unsuitability with plastics and high volume changes. On the other hand, non-paraffins generally include esters, glycols and fatty acids. These materials generally have different characteristics which offer choices to the user. The drawback of non-paraffins is their flammability, thus, limiting their application to lower temperature ranges. However, building applications with fatty acids as PCMs are gaining popularity due to the high latent heat, low super cooling characteristics, no phase separation, good thermal reliability and thermal stability [6-10]. As for the eutectic PCMs, these are a mixture of two or more PCMs in order to tailor the material according to the requirement. This kind of materials is applied when certain characteristics like specific melting points cannot be obtained with other PCMs [11].The choice of PCMs for applications is determined based on the following factors [12-14].

(1) *Latent heat thermal energy storage (LHTES):* The large latent heat capacity and the convenient phase change temperature;
(2) *Other thermo-physical properties:* Suitable heat capacity, high thermal conductivity, appropriate density, little volume change, no phase segregation and small super-cooling degree;
(3) *Chemical properties:* Corrosion-resistant, chemically stable and non-toxic, compatible with container, non-flammable and environmentally friendly;
(4) *Economic requirements:* low cost, disposal and recyclability.

The main PCM drawbacks include [15,16]:

(1) *Super-cooling:* During the cooling process, most PCMs take time for solidification after getting below the phase change temperature; a temperature difference is needed to start PCM nucleation and subsequent crystallization. In the case of organic PCMs, super-cooling

temperature difference is small (few degrees), while it can be large for water- and salt-based PCMs (tens of degrees).

(2) *Phase separation:* PCMs tend to be pure with almost no phase separation. However, it is not the case for non-pure PCMs, like some eutectic PCMs and salt hydrate. Two or more different layers may separate during the melting process because of their density difference, and the PCM homogeneity may not be reversible.

(3) *Low thermal conductivity:* The thermal conductivity of most organic PCMs is less than 0.4 W/(m.K), while that of most inorganic PCMs is less than 1.0 W/(m.K). The low thermal conductivity of pure PCMs causes a slow-down of heat transfer rate, resulting in slow heat storage/release in TES applications, besides the difficulty to maintain a constant temperature in thermal management systems.

6.2 Thermally Energy-efficient Construction Materials

Energy consumption is on a constant rise as a consequence of rapid population growth along with improving living standards. Researchers, over the years, have tried to find the solution for this rising concern with different technologies. As a result, the idea of incorporating phase change material into different building elements gained importance. It has been reported in the literature that including PCMs in building walls will decrease the mismatch between supply and demand of electricity [14]. PCMs are expected to decrease the temperature fluctuations in the buildings, thereby, enhancing the indoor thermal comfort. Using PCMs in buildings during winter will help inefficient utilization of solar energy. In summer, it helps in decreasing internal cooling loads.

If a building is viewed as a thermodynamic system, energy conservation across the building wall implies balance of energy generation within the building and energy accumulated in the wall with PCM and energy transferred across the wall. This balance depends upon LHTES properties of PCM, the quantity of PCM and resistance to energy flow offered by PCM across the wall.

Another important reason for inclusion of PCMs in buildings is to decrease the use of heating and cooling devices in buildings during the peak hours of the day. As reported in literature, the main reason for choosing different PCMs among various thermal energy storage technologies is the storage of 5-14 times more energy in PCMs than sensible heat storage material for the same volume of these materials [16].

Inappropriate PCMs, if used for building applications, might result in inefficient charging and discharging cycles of PCM. The melting point of PCMs must lie within the temperature range of 18-40 °C, so as they may be deemed appropriate for building applications [17].

PCMs can be included in construction materials through direct mixing or dipping construction materials in liquid PCM or mixing PCM encapsulated micro and macro capsules into the materials. Direct mixing involves mixing PCMs and construction materials in certain proportions. In the dipping method, the porous substrate of building element is dipped into PCMs for absorption. Encapsulation involves covering PCM with shells to prevent their leakage during phase change process. PCMs can be encapsulated in either micro or macro scale.

Micro encapsulation involves covering PCM particles of micron size with polymer shells. Macro encapsulation involves confining the PCM in a container appropriate to the construction process and element being constructed. The form-stable composite PCMs (FS-CPCMs) are other viable alternatives for inclusion into construction materials when the thermal stability issue arises [18].

6.2.1 Incorporation of PCM into Concrete

Generally, there are a few techniques used to incorporate PCMs in concrete. Three of the widely used methods are [19]:

(i) *Immersion*: Liquid PCM is simply immersed into porous concrete [20].
(ii) *Impregnation*: Liquid PCM is impregnated into porous aggregate by vacuum force [21,22].
(iii) *Direct mixing*: Encapsulated PCM is directly mixed with the concrete mix during mixing stage [23].

The thermal and mechanical properties of PCM-concrete composite are highly affected by the method of PCM impregnation into concrete, as reported in previous works [21-25].

6.2.2 Overview of Clay-based Composite PCMs

The porous structure and the relatively high surface area mark the nature of the majority of clay mineral materials. Hence, a compound of clay mineral and PCM can be fabricated due to the absorption ability. On the other hand, clay mineral materials that have small interlayer spacing and low specific surface area are inconvenient to adsorb liquid PCM. The clay mineral materials interact with PCM through capillary force, along with, van der Waals force, hydrogen bonding and surface tension [26,27]. These interactions can help limit PCM leakage in a clay mineral-based F-SPCM. Clay mineral-based F-SPCM are commonly based on kaolin [28], vermiculite [29,30], sepiolite [31], bentonite [32], perlite [29,30,33], SiO_2 [34], attapulgite [35], diatomite [29,30], fly ash [36] and opal [37], etc. The interlayer spacing of kaolin is small, which make it tough to compound with PCM. Sepiolite, vermiculite, expanded perlite and attapulgite have high specific surface area. Three-dimension crystal structure of SiO_2 also effectively prevents PCM leakage. Fly ash and diatomite have significant porosity. Montmorillonite has good cation exchange capacity and prominent specific surface area. Moreover, the frequently used organic PCM with clay mineral materials comprise fatty acids [38], alkanes [39], polymers [40] and the binary/ternary eutectic mixtures based on these [41,42]. There are many methods to prepare clay mineral-based FS-CPCM such as melting adsorption [31], vacuum impregnation [43] and melting intercalation [44]. Clay mineral-based FS-CPCMs have numerous qualities and benefit from abundance and moderate price of clay minerals.

Kaolin-based FS-CPCM

The thermal properties of FS-CPCM are highly affected by the different interactions and porous structure of kaolin [45,46]. Paraffin incorporated kaolin was prepared by Li *et al.* [47]. The FS-CPCM exhibited different latent heat capacities depending on the mass fraction of paraffin. An increasing enthalpy of composite is found with the increasing mass fraction. Other kinds of PCMs are also used for the preparation of FS-CPCM, such as binary PCMs, which are binary eutectic mixture prepared in precise ratio. Binary systems have low phase change temperature in comparison with single PCM. The capric acid-lauric acid (CA-LA) eutectic mixture was incorporated into kaolin by Liu *et al.* [43] to generate a FS-CPCM. Thermal characterization revealed that lauric acid, capric acid and their eutectic mixture have the respective phase change temperature of 44.16 °C, 30.97 °C and 18.96 °C, along with a small reduction of the latent heat capacity of CA-LA eutectic mixture. Furthermore, a small change of phase change temperature of the eutectic mixture was reported due to kaolin restriction [44]. In other studies, thermal stability of kaolin-based FS-CPCM was reported by Song *et al.* [45] and Sari [46]. A relative increase in PCM's onset degradation temperature was observed, which confirms that the composite is thermally stable.

The commercial use of kaolin-based FS-CPCM dictates high thermal reliability along with good chemical reliability. To quantify these properties, thermal cycling tests are applied on the composite. A slight change was observed, therefore, FS-CPCM is considered to be thermally and chemically stable [46,47].

Diatomite-based FS-CPCM

Diatomite has also been a subject of FS-CPCM. The paraffin/diatomite composite was reported by Li *et al.* [39]. Diatomite with three different particle sizes was tested: DP1, DP2 and DP3. The particle size analysis showed that DP1 had the finest particle size (90% was less 11.662 μm), while DP2 and DP3 had similar distribution (few hundred micrometers). DP1-based FS-CPCM exhibited higher enthalpy in comparison to DP2 and DP3 as a result of important difference in loaded quantity of paraffin, which was 47.9%, 35.8% and 34.4% for DP1, DP2 and DP3-based FSPCM, respectively. Slight change was noticed in the phase change temperature of the composites. In another work, Lei *et al.* [48] fabricated a FS-CPCM based on NA-P/diatomite. Different mass proportions of paraffin and $Na_2SO_3 \cdot 10H_2O$ affected the latent heat capacity of the composite significantly. In fact, differential scanning calorimetry (DSC) curves revealed two endothermic peaks due to different melting points. The more paraffin mass ratio was incorporated, the more latent heat capacity of the composite was enhanced. However, small change in phase change temperature was observed with change in paraffin mass ratio. Diatomite based FS-CPCM was also synthesized by Qian *et al.* [40]; three different PCMs were used, PEG as a low temperature PCM, $LiNO_3$ as middle temperature PCM and Na_2SO_4 as a high temperature PCM. The three diatomite-based FS-CPCMs had the respective phase change temperatures of 887.61 °C, 57.92 °C and 250.70 °C. The high absorption ratio of PCMs was observed to enhance the enthalpy of composite.

The lauric acid/diatomite FS-CPCM was investigated by Fu *et al.*

[49] for its thermal stability. At temperatures below 146 °C, weight loss was insignificant. Up to 189 °C, only 0.61% weight loss was observed. Jeong *et al.* [50] reported similar results for PA-CA/diatomite FSPCM. Up to 100 °C, no significant degradation was noticed. Thermogravimetry (TG) curves showed degradation starting around 100 °C and ending at 250 °C, which indicated good thermal stability [51].

Thermal cycling tests help evaluate the thermal reliability. Karaman *et al.* [52] verified the thermal reliability of PEG/diatomite. 1000 thermal cycling tests revealed a change of 1.94 °C and 0.41 °C in melting and freezing temperature, respectively. Only 1.1% of change was observed in latent heat capacity, indicating excellent thermal reliability.

Sepiolite-based FS-CPCM

Sepiolite-based FS-CPCM have been fabricated by Jiang *et al.* [53,54] using stearic acid and lauric acid as PCMs. The two types of sepiolite-based composites exhibited the respective temperature range of phase change of 52.20-58.70 °C and 43.82-46.48 °C. This was slightly below the phase change temperatures because of the weak interactions between the carrier and PCM. Jia *et al.* [55] prepared sepiolite-based FS-CPCM using paraffin (P), CA and CA-P eutectic mixture. The CA-P had lower phase change temperature than that its base PCMs, but rather similar latent heat capacity.

Wu *et al.* [56] fabricated neopentyl glycol/sepiolite composite as FS-CPCM. A weight loss of 20.77% was measured below 83.86 °C, which indicated a relatively low thermal stability of NG/sepiolite. LA-SA/sepiolite composite was reported by Deng *et al.* [31]. 1.4% of weight loss of the composite was measured below 100 °C. In other works, Jiang *et al.* [53,54] and Guo *et al.* [57] reported the thermogravimetric analysis of sepiolite-based composites. The weight loss was negligible, thus, the composites had an excellent thermal stability.

Montmorillonite-based FS-CPCM

Montmorillonite (MMT) based FS-CPCM has weak diversity in phase change temperature, which is attributed to the interaction between MMT and PCM. The composite has a lower melting temperature compared to the PCM because of the strong interactions [58,59]. The mass fraction of the PCM and its enthalpy dictate the latent heat capacity of the composite [60]. It increases gradually as a function of PCM mass fraction [61]. The phase change range of the binary eutectic mixture is lower than the constituent single PCMs [62]. In another study, neopentyl glycol was used as PCM to fabricate MMT-based FS-CPCM by Shen *et al.* [63]. The DSC analysis revealed two adsorption peaks, one linked with solid-solid enthalpy and the other attributed to the solid-liquid phase transition of neopentyl glycol. The first occurred at 47.3 °C with 128.1 J/g, whereas the second one was observed at 132.5 °C with 44.83 J/g. The same phenomenon was recorded for butyl stearate with the solid-liquid phase transition as the main phase change process.

Perlite-based FS-CPCM

The diversity in paraffins' phase change ranges implies different phase change temperatures of expanded perlite-based FS-CPCMs [64-68]. Lu *et al.* [64] reported the investigation of paraffin/expanded perlite composite, with different weight ratios of paraffin. Respectively, 20 wt%, 40 wt%, 60 wt% and 80 wt% of paraffin were loaded into EP. The thermal characterization showed that the latent heat capacity of paraffin/EP FS-CPCM enhanced with the incremental mass ratio of paraffin. The fatty acid/EP composite was fabricated as FS-CPCM by Shang *et al.* [69]. Melting dilation of fatty acid causes the fatty acid/EP to have a little higher phase change temperature than that of pure PCM. On the other hand, Sari and Karaipekli [70] investigated EP-based composite with capric acid. The composite's phase change temperature was found to be marginally lower than that of pure CA, as EP is low-alkaline clay. However, the restriction of EP's poor structure widened the phase change range of the composite compared to the pure PCM [69]. The binary eutectic mixture tends to have lower phase change temperature compared to the single PCMs. Additionally, ternary eutectic mixtures have even lower phase change temperature that of any binary eutectic mixture [71,72]. Yu *et al.* [73] reported the investigation of EP-based FSPCMs based on the binary/ternary mixture of hexadecanol-PA (7:3), LA-PA (76:24), hexadecanol-PA-LA (5(7:3):5) and LA-PA-MA (60.8:19.2:20). Their respective phase change temperatures were observed to be 44.7 °C, 31.7 °C, 27 °C and 26.9 °C.

The use and development of EP-based FS-CPCM depend on the thermal reliability of such material. The thermal reliability of paraffin/EP was tested by Lu *et al.* [74]. Around 59.4% of paraffin mass fraction was achieved. The thermal reliability of paraffin/EP FSPCM was also tested by carrying out 1000, 3000 and 5000 thermal cycling [75]. A slight change was noticed in the FTIR absorption peak and thermal properties, which indicated a good thermal reliability of paraffin/EPFS-CPCM.

SiO_2-based FS-CPCM

SiO_2 composites have also been studied for various applications. Wang *et al.* [76] fabricated SiO_2-based composite using PEG10000, PEG6000 and PEG4000. The respective latent heat capacities of each composite changed by 3.5%, 5.5% and 6.8% after 1000 thermal cycling tests, which are satisfactory. The stearic acid/silica fume composite was observed to have thermal reliability after 600 thermal cycling between 30-80 °C [77].

Vermiculite-based FS-CPCM

PCM leakage has been a major challenge requiring appropriate resolution. Use of vermiculite as base clay for PCM composites has demonstrated a leakage prevention potential. Many researchers have investigated vermiculite-based FSPCM with organic PCMs, mainly, unitary organic PCMs like paraffin [78] and n-octadecane [79], along with binary organic PCMs such as fatty acids eutectic mixtures [79,80] and ternary eutectic PCM [81]. Zhang *et al.* [82] reported the fabrication of a vermiculite-based FS-CPCM using LA-PA-SA eutectic mixture. A slight decrease in latent heat capacity was measured after 1000 cycling tests, which proved the thermal reliability of the composite. The different fatty acid eutectic mixtures were tested by carrying out 3000 and 5000 thermal cycling tests, which exhibited a good thermal reliability of the composites [80,82]. In another study, n-octadecane/vermiculite FS-CPCM was fabricated by Chang *et al.* [79]. The FS-CPCM was observed to lose negligible weight at working temperature, which indicated its excellent thermal durability performance.

Fly Ash (FA)-based FS-CPCM

Fly ash was studied by Jiang *et al.* [83] as base clay of lauric acid FS-CPCM using melting impregnation method. The composite's phase change range was found to be 44.31-46.01 °C. Melting temperature enhancement was caused by the strong interactions between PCM and FA, while the weak interactions cause the melting temperature to decrease [84]. Yang *et al.* [36] reported the preparation of cetyl alcohol-decylic acid/FA FS-CPCM. The composite had a phase change temperature of 25.7 °C and a latent heat capacity of 70.51 J/g. 48% PCM mass fraction was achieved in the thermal studies. Furthermore, Hua *et al.* [85] used aluminum (Al) as a PCM and FA as a carrier. The goal was to investigate the feasibility of clay mineral-based PCM composite at high temperature. Similarly, Wang *et al.* [86] studied Al/FA composite. It was found that the composite had an endothermic peak at 579 °C; the magnetic separation affected the latent heat capacity significantly, by decreasing it from 35.65 J/g to 20.29 J/g.

Attapulgite-based FS PCMs

Attapulgite (ATP) is an abundant low cost clay mineral with a layer chain-like structure and hydrated magnesium aluminum silicate composition. Recently, attapulgite has attracted significant attention in diverse research fields due to its interesting properties, such as natural nano-channel structure, high specific surface area, chemical composition and ability of exchange properties involving large organic molecules [87-94]. PCM leakage during phase change process can be prevented due to surface tension and capillary force between organic PCM and macro-size pores of attapulgite. Additionally, the optimal thermal properties of attapulgite-based FSPCM result in many potential applications, which has attracted many researchers to study organic PCM loading into attapulgite. The studied organic PCMs are either unitary like paraffins [46,47] and n-octadecane [89] or binary eutectic mixtures as SA-CA [35]. The diverse methods were reported to prepare the attapulgite-based F-SCPCM. The SA-CA/attapulgite F-SCPCM was prepared using vacuum fusion absorption method [35] and vacuum method [90] was used for the preparation of CA-PA/attapulgite F-SCPCM. Vacuum impregnation method [89], heat fusion method [90] and physical adsorption method [91] were also used for the preparation of attapulgite-based FS-CPCM. The n-octadecane/attapulgite composite was prepared by Chen *et al.* [89]. An irregular change in phase change range was observed as a function of PCM mass fraction. Interactions between attapulgite and n-octadecane molecules might have caused such an effect on phase change range. With the incremental mass ratio of PCM, the latent heat capacity of the composites increased. Liang *et al.* [92] reported the fabrication of FS-CPCM based on spongy attapulgite (S-attapulgite) and purified attapulgite (P-attapulgite), along with carboxylic acid as PCM. The melting temperature of S-attapulgite F-SCPCM was slightly higher than pure acid because of the melting dilation of the acid in S-attapulgite's porous structure. Phase change range of acid/P-attapulgite and acid/S-attapulgite stayed unchanged. On the other hand, there was a difference in the latent heat capacity of the two composites, obviously due to the different weight ratio of PCM. In general, the latent heat capacity of P-attapulgite based FS-CPCM was higher than the other.

As seen from the literature survey, the studies about the attapulgite based-composite PCMs are limited in number. Therefore, there is a need to widen the variety of novel composite PCMs with attapulgite exhibiting high LHTES efficiency. Besides, palmitic acid (PA)-stearic acid (SA) are suitable organic PCMs for the impregnation with attapulgite clay for passive solar TES applications depending on the climatic conditions. To the best of our knowledge, no investigation has been reported in the literature on the fabrication, chemical-morphological characterization and LHTES property determination of attapulgite/(PA-SA) composites, thus, these composites have been studied in detail in the current study. In the composites, the prepared eutectic fatty acids have LHTES function while attapulgite acts as supporting matrix in the composites. The specific surface area of the attapulgite was determined using Brunauer-Emmet-Teller (BET) method. XRD analysis, scanning electronic microscope (SEM) with energy dispersive X-ray spectroscopy (EDX) and Fourier transformation infrared spectroscope (FTIR) techniques allowed to characterize morphologically and chemically the attapulgite based-building composite PCMs. The LHTES properties and thermal reliability of the prepared composite PCMs were analyzed by DSC.

6.3 Experimental

6.3.1 Materials

PA and SA were obtained from Sigma-Aldrich (Germany). Attapulgite was obtained from Karaman mines Company (Turkey). According to the manufacturer, it had weight composition of 83.4%

SiO_2, 5.60% Al_2O_3, 1.0% CaO, 1.0% Fe_2O_3 and 0.7% of other mineral oxides (K_2O, Na_2O and TiO_2). The clay sample was sieved through 150-mesh sieve. It was dried at 105 °C for 24 h before use.

6.3.2 Preparation of Fatty Acid Eutectic Mixture (EM)

For the preparation and characterization of the PA-SA eutectic mixture (EM), the following process steps were followed:

(i) Firstly, the eutectic composition ratios between the components were theoretically calculated by Schroder equation given as [94]:

$$\ln X = \frac{\Delta H m}{R}\left(\frac{1}{Tm} - \frac{1}{T}\right)$$

where, X is the mole fraction of the major component at temperature T, R the gas constant 8.314 J/mol K, ΔH_{m} the molar enthalpy of fusion and T_m is the melting temperature of the pure component.
(ii) The mixtures were prepared at different composition percentages above and below the regarded theoretical ratio.
(iii) DSC thermograms of each mixture prepared at different percent compositions were recorded at the same heating rates. The dual peak or the shoulder-peak in the thermogram showed non-eutectic binary mixture as the mixture with eutectic combination is represented by single peak-thermogram.
(iv) The phase change temperature and phase change enthalpies of the eutectic mixture were determined by methods related to the DSC analysis technique.

6.3.3 Preparation of Attapulgite/PA-SA as FS-CPCM

In the first stage of the experimental procedure, the attapulgite/PA-SA composites were prepared by vacuum infiltration process. The experimental procedure is shown schematically in Figure 6.1.

As seen from the scheme, the specified amount of attapulgite was firstly put inside a flask connected to a vacuum pump. The vacuum operation was maintained at 0.18 MPa for 30 min as EM in liquid state was dropped gradually to the flask using a funnel. Subsequently, the vacuum was released and air was allowed for 30 min to make easy infiltration. The prepared attapulgite/PA-SA composite was cooled up to 10 °C to guarantee entire crystallization of EM into the porous network of attapulgite. This procedure was replicated for different weight combinations ranging from 15 to 40%.

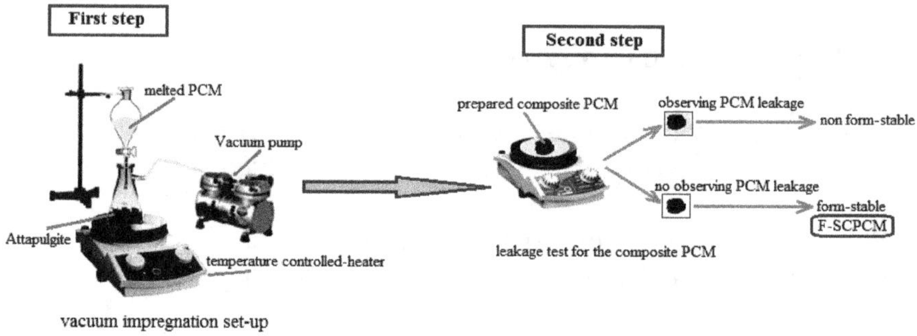

Figure 6.1 The schematic of the experimental procedure used for the preparation of attapulgite/PA-SA as FS-CPCM.

In the second stage, the optimal infiltration ratio was established by adopting leakage test. The prepared composite was placed onto a filter paper and heated above the melting temperature of EM. The leakage of attapulgite/PA-SA composite was checked by controlling the filter paper through the heating treatment.

6.3.4 Characterization of Attapulgite/PA-SA as FS-CPCM

The structural properties and surface area of the attapulgite clay used in this work was determined by BET method. The specific surface area was measured by gas adsorption technique, which is a measure of the exposed surface of a material reported in terms of square meters per gram. Micromeritics Gemini VII 2390 V1.03 surface analyzer was used to perform the analysis (Figure 6.2).

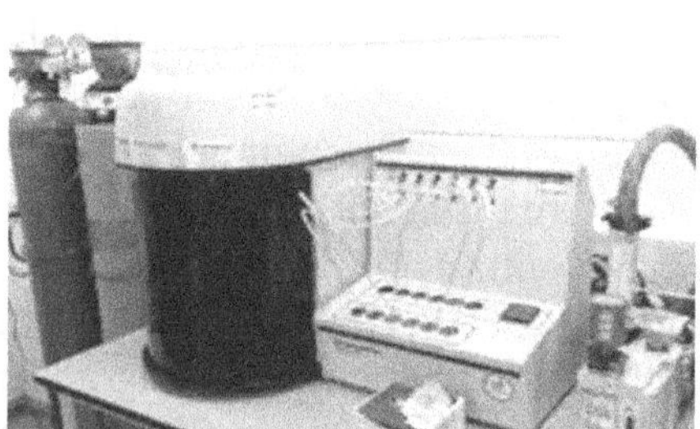

Figure 6.2 Micromeritics Gemini VII 2390 V1.03 model surface analysis instrument.

The chemical characterization of attapulgite and the developed attapulgite/PA-SA composite were performed by using Jasco 430 FT-IR spectrometer (Figure 6.3). The sample was prepared by mixing with KBr at a ratio of 1:100 and a very thin pellet was subsequently formed under 13 ton pressure. The spectrum was obtained at wavenumber range between 4000 and 650 cm^{-1} with a resolution of 4 cm^{-1}.

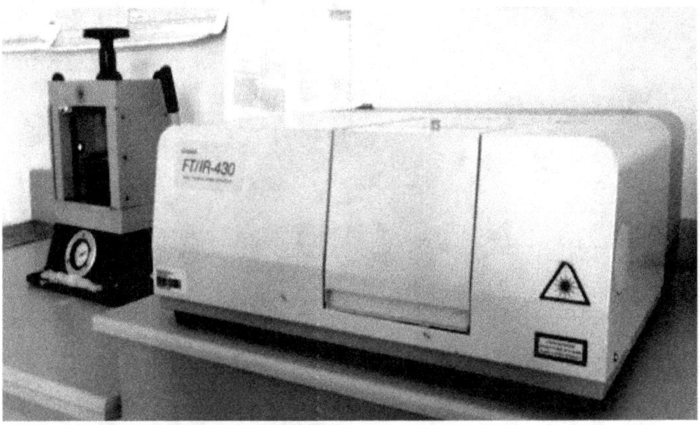

Figure 6.3 JASCO 430 model FT-IR instrument.

The microstructure of the fabricated FS-CPCM was examined using a LEO 440 SEM equipped with EDX (Figure 6.4). Before SEM/EDX analysis, a small amount of sample was adhered on a copper SEM stub by conductive adhesive and gold-coated in a high vacuum evaporation coating machine.

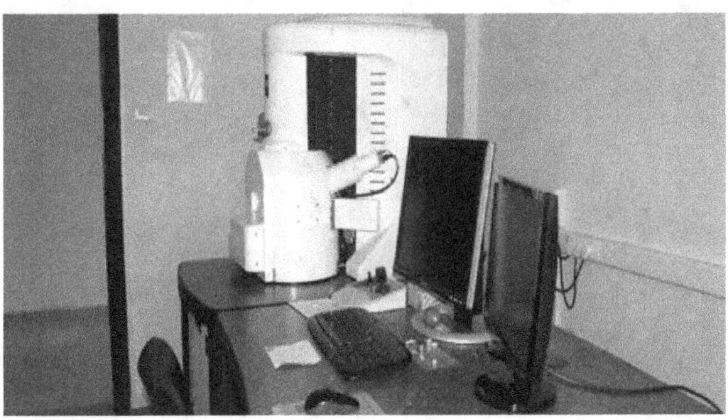

Figure 6.4 LEO 430 model SEM instrument.

X-ray diffraction (XRD) analysis of the produced attapulgite/PA-SA composite was conducted using a PANalytical X'-Pert[3] powder diffractometer (45 kv, 40 mA) with Cu (K_α = 1.5406 Å) irradiation in the Bragg angles range of 2θ ($0 \leq 2\theta \leq 70°$) at a step size of 0.0131° (Figure 6.5).

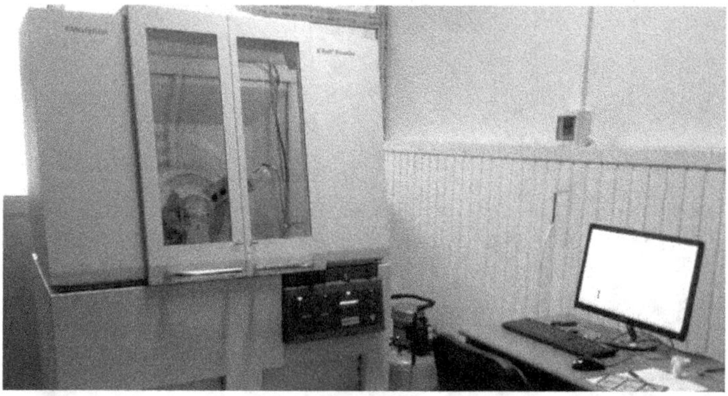

Figure 6.5 PANalytical X'-Pert3 model XRD instrument.

The LHTES properties of EM and the fabricated FS-CPCM were measured at the same heating/cooling rate of 3 °C min^{-1} by DSC technique (Perkin Elmer-JADE model). The measurements were repeated three times and the accuracy in the temperature and enthalpy data was calculated as ±0.012 °C and ±0.59 J/g, respectively. Figure 6.6 shows the DSC instrument used in the measurement of the LHTES properties of EM and the prepared F-SCPCMs.

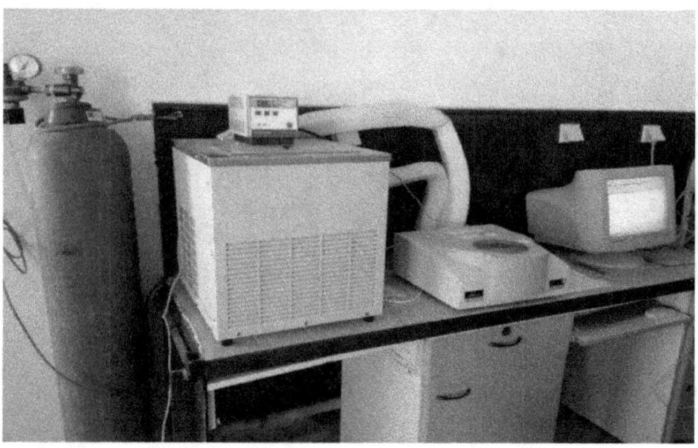

Figure 6.6 Perkin Elmer-JADE model DSC instrument.

By using a thermal cycler (BIOER TC-25/H), the developed FS-CPCM was subjected to a thermal cycling test including 500 melting/freezing cycles. After 500 cycling treatments, the DSC analyses were repeated under the same analysis conditions applied for the uncycled composite samples to determine the possible changes in the LHTES properties after the cycling test. To obtain information about the chemical stability of FS-CPCM, the FT-IR spectra taken before and after the cycling test were also compared. Figure 6.7 shows the image of the thermal cycler used for the thermal cycling tests.

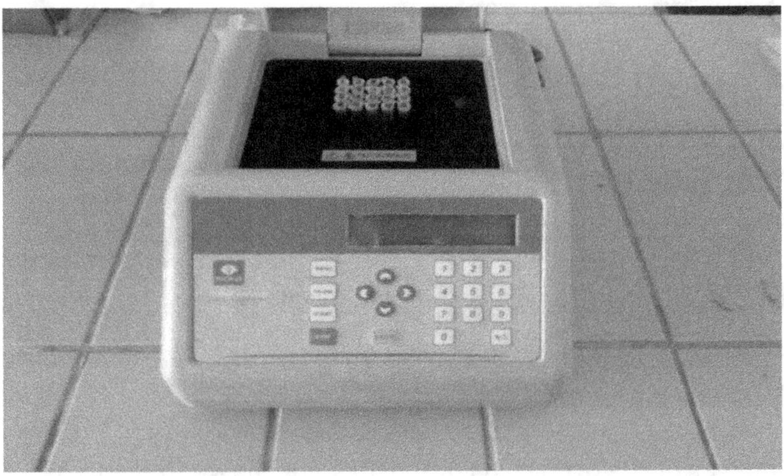

Figure 6.7 BIOER-TC-25/H model thermal cycler used for the thermal cycling tests.

6.4 Results and Discussion

6.4.1 Physico-chemical Properties

BET

Figure 6.8 shows the BET results including the structural properties, surface area and nitrogen gas adsorption diagram of attapulgite clay.

As seen from the results, attapulgite had a large surface area of 114.21 m^2/g and single point adsorption total volume of pores of 0.312 cm^3/g.

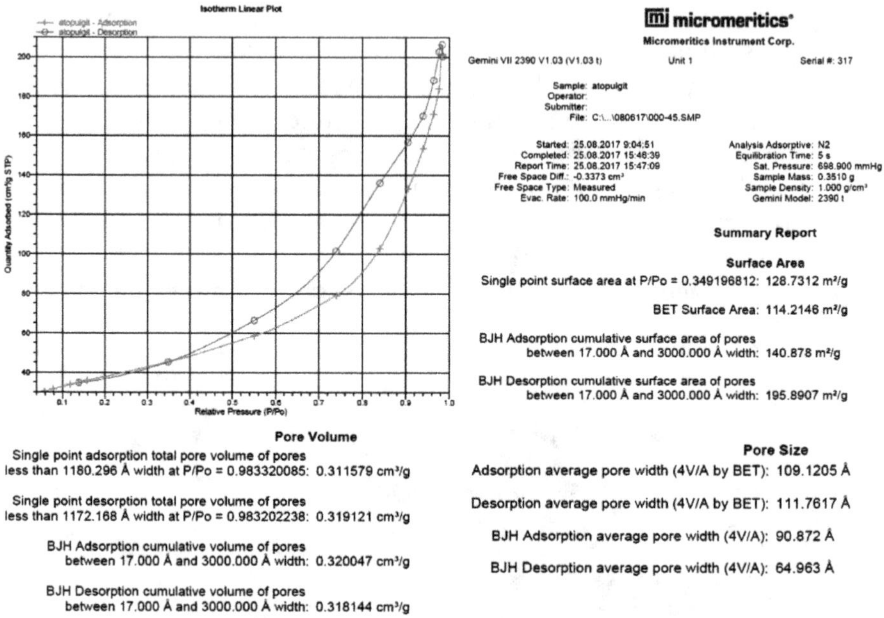

Figure 6.8 BET analysis results of the attapulgite clay.

Leakage Test Results

The images of the prepared composites are shown in Figure 6.9. The maximum absorption of PA-SA eutectic mixture by attapulgite was 40 wt% without any observed melted leakage. Thus, the composites with these combinations were termed as FS-CPCM.

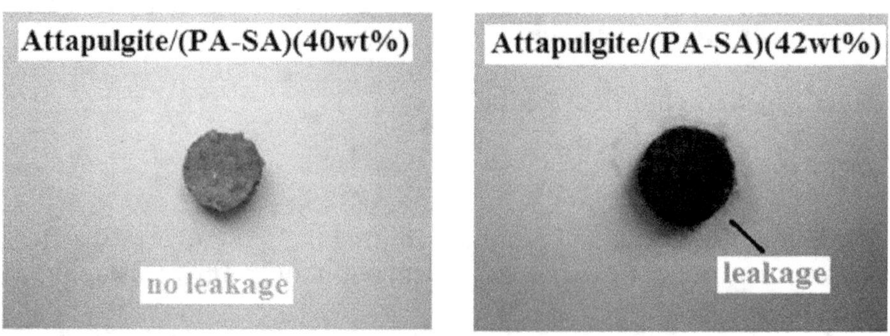

Figure 6.9 The photographs of the prepared composite PCMs including EM above/below the form-stable combination ratio.

SEM/EDX

The microstructures of attapulgite and the prepared FS-CPCM are shown in Figure 6.10. As seen from the SEM photographs, the surface of attapulgite consisted of arbitrary-shaped particles with needle-like structure. As also seen from the SEM image of the FS-CPCM, EM was homogenously dispersed throughout the surface. Although the electron beam was applied to the composite sample under high voltage, no liquid phase was detected as a second phase. It was due to the structural resistance of cement and capillary/surface tension forces in the matrix. These results proved the existence of excellent physical compatibility among the components of the FS-CPCM.

Figure 6.10 The microstructures of the attapulgite and the prepared FS-CPCM.

The EDX mapping and elemental analysis results of attapulgite as well as attapulgite/(PA-SA) composite are shown in Figure 6.11 and

6.12. Attapulgite consisted mainly of Si, Al, Mg, Ca, Fe, O and K. The distribution of these elements was observed to be homogenous on the surface.

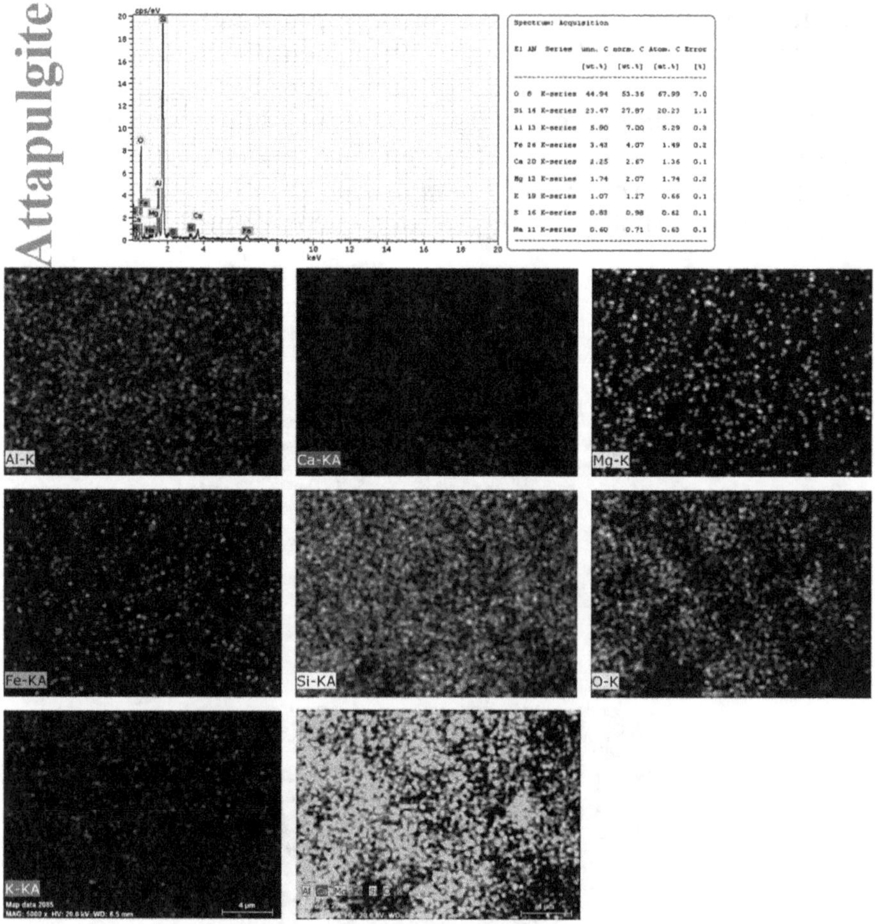

Figure 6.11 EDX mapping and elemental analysis results of attapulgite.

FTIR

The FTIR analysis was performed to describe the probable specific interactions between attapulgite and PA-SA eutectic mixture (Figure 6.13 and 6.14). As seen from the FT-IR spectrum of attapulgite in

Figure 6.13, the band in the range of 3200-3750 cm^{-1} resulted due to the asymmetrical stretching of hydrogen bonded OH groups due to H-O-H ingredient and Si-OH group, while the bending vibration of these groups was detected in the range of 1605 cm^{-1}. The bands in the range of 1090 cm^{-1} and 478-950 cm^{-1} were characterized as the stretching vibrations of Si-O-Si and bending vibrations of Si-O, Mg-O and Si-O-Al groups.

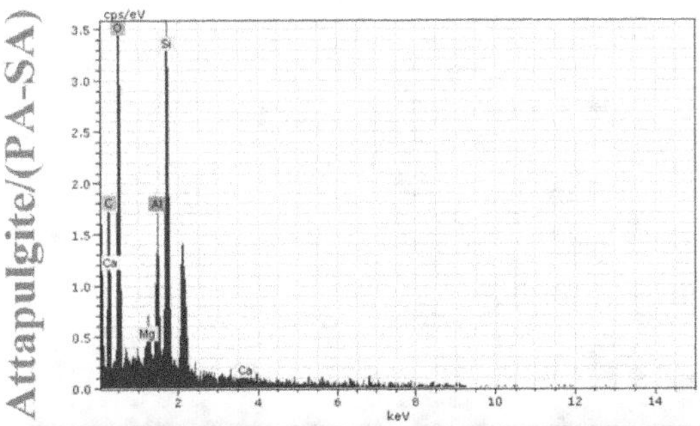

El	AN	Series	unn. C [wt.%]	norm. C [wt.%]	Atom. C [at.%]	Error [%]
O	8	K-series	62.71	62.72	62.94	21.9
C	6	K-series	21.42	21.42	28.64	9.8
Si	14	K-series	10.53	10.53	6.02	0.5
Fe	26	K-series	1.96	1.96	0.56	0.2
Al	13	K-series	1.93	1.93	1.15	0.1
Ca	20	K-series	0.74	0.74	0.30	0.1
Mg	12	K-series	0.41	0.41	0.27	0.1
K	19	K-series	0.30	0.30	0.12	0.1
		Total:	100.00	100.00	100.00	

Figure 6.12 EDX mapping and elemental analysis results of attapulgite/(PA-SA) composite.

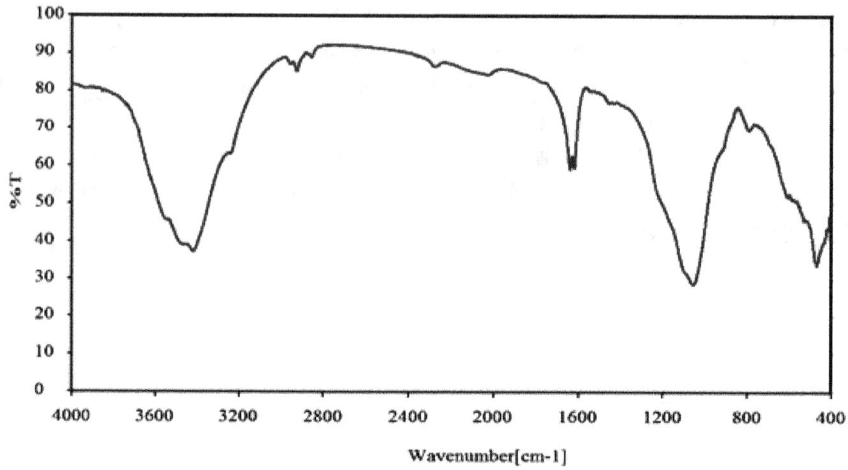

Figure 6.13 The FT-IR spectrum of attapulgite.

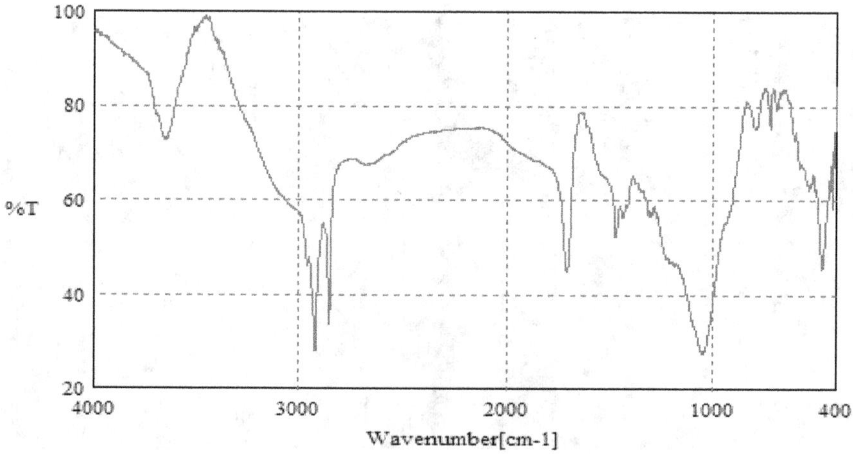

Figure 6.14 The FT-IR spectrum of the prepared FS-CPCM.

In case of the composite, the stretching and bending vibrations of Si-OH bands were seen at 3200-3700 cm^{-1} and 1046 cm^{-1}, while the bending vibrations of Si-O, Mg-O and Si-O-Al groups were observed at 450-520 cm^{-1}. These bands confirmed the presence of the attapulgite component of the composite PCMs. Moreover, in the FT-IR spectrum of the eutectic mixtures, the bands at 3100-3600 cm^{-1} were regarded as the stretching vibration of OH group (as the bands at 933 cm^{-1} and 732 cm^{-1} were due to the bending vibration of this group). The asymmetrical and symmetrical stretching bands at 2980-2850 cm^{-1} and 1470 cm^{-1} were attributed to the presence of alkyl groups (-CH_2 and -CH_3). The band at 1704 cm^{-1} was identified as the stretching band of C=O. The stretching band of C-O groups was over layered by bending vibrations of Si-O, Mg-O and Si-O-Al groups. Additionally, no new peak was observed in the spectrum. This confirmed that no chemical reaction took place between the components of FS-CPCM. However, small shifts observed in some characteristic bands of the FS-CPCM could still be attributed to the probable intermolecular physical attractions between the characteristics groups of EM and attapulgite.

XRD

The XRD patterns of the attapulgite/EM composites are shown in Figure 6.15. The XRD pattern of attapulgite included the crystal diffractions at °2θ values of 26.67°, 28.02°, 35.56° and 39.50°. The

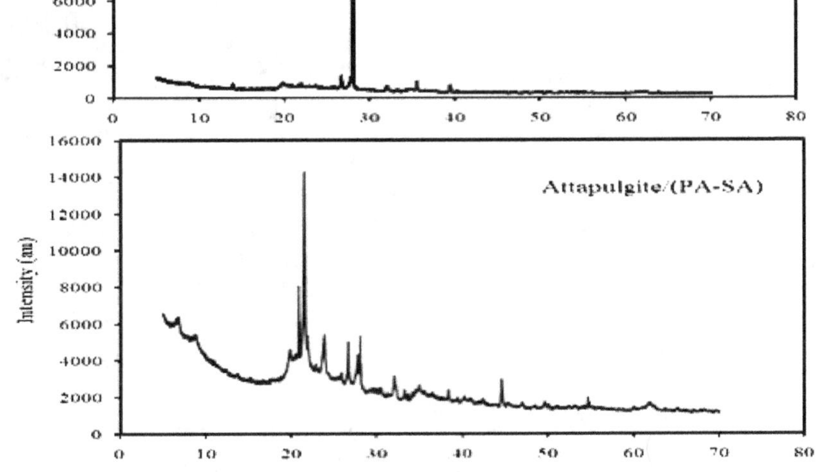

Figure 6.15 The XRD results of the attapulgite and the prepared FS-CPCM.

attapulgite/(PA-SA) composite had the diffraction peaks at 19.74°, 21.52°, 23.78°, 26.61°, 26.63°, 27, 58°, 28.06°, 32.04°, 34.86°, 44.59° and 39.54°.

As seen from these findings, the XRD patterns of the attapulgite/EM composites prepared as FS-CPCM contained all crystal diffraction peaks of its pure components. Moreover, some of the peaks overlapped because of the °2θ values close to each other. On the other hand, little shifts in °2θ values and peak intensities of EM were observed. It indicated that its crystal structure was not disturbed after impregnation process and the observed shifts might be due to the capillary and tension forces.

6.4.2 LHTES Characteristics of FS-CPCM

DSC Analysis of the Prepared EM

The DSC findings of the PA-SA mixtures are demonstrated in Figure 6.16 and Table 6.1. In each of the DSC thermograms, it was possible to observe the shouldered-peaks of each mixture, which indicated non-eutectic composition. On the other hand, the eutectic mixture behaved as a single substance (pure substance), showing regular phase change at a lower temperature than the melting temperature of the each fatty acid component.

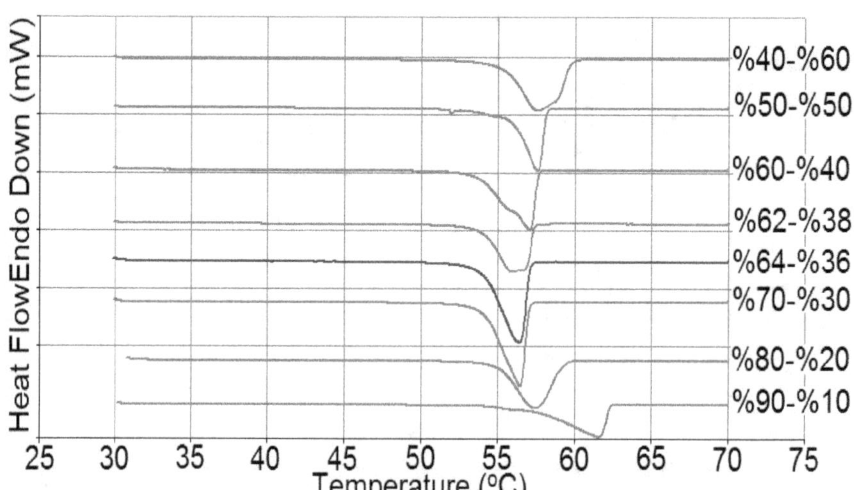

Figure 6.16 DSC thermograms of PA-SA mixtures prepared using different weight ratios.

By taking account these considerations in the DSC thermograms, the eutectic weight composition was determined as 64-36 wt% for PA-SA mixture. Moreover, the eutectic melting temperature was measured as 54.08 °C, while the latent heat capacity was found to be 187.59 J/g.

Table 6.1 The LHTES properties of the prepared FS-CPCM and its components

PCM	Melting temperature (°C)	Solidification temperature (°C)	Latent heat of fusion (J/g)	Latent heat of solidification (J/g)
Palmitic acid (PA) based-FS-CPCM	62.15	61.97	220.25	-216.89
Stearic acid (SA) based-FS-CPCM	69.64	68.98	232.03	-229.35
PA-SA (64-36 wt%) based-FS-CPCM	54.08	53.93	187.59	-185.27
Attapulgite/(PA-SA) based-FS-CPCM	53.26	53.83	73.64	-72.14

Figure 6.17 demonstrates the DSC thermogram of the fabricated attapulgite/(PA-SA) composite with combination of 40 wt%. As clearly observed, the prepared form-stable attapulgite/(PA-SA) composites melted at 53.26 °C and solidified at 53.83 °C. Moreover, the LHTES capacities for melting and freezing periods were measured to be 73.64 J/g and -72.14 J/g, respectively. On the other hand, when compared to EM, the phase change temperatures of the FS-CPCM decreased by -0.82 °C for melting and -0.10 °C for freezing processes. These could be attributed to the physical interactions between -OH group of EM and alkali groups (Si-OH, Al-Si-O, Mg-O, Ca-O, etc.) of attapulgite, which was proved through the FTIR findings. Although little deviations were observed in the phase change temperatures, the FS-CPCM could be evaluated as fitting LHTES material to regulate indoor temperature of building envelopes. Additionally, its LHTES capacity of melting was slightly lower than the corresponding theoretical values (75.03 J/g) calculated by multiplying the latent heat capacities of EM with its impregnation ratios in the composite. However, it can be deduced from Table 6.2 that the produced FS-CPCM had sufficient LHTES capacities when compared with attapulgite-based composite PCMs reported in literature [93].

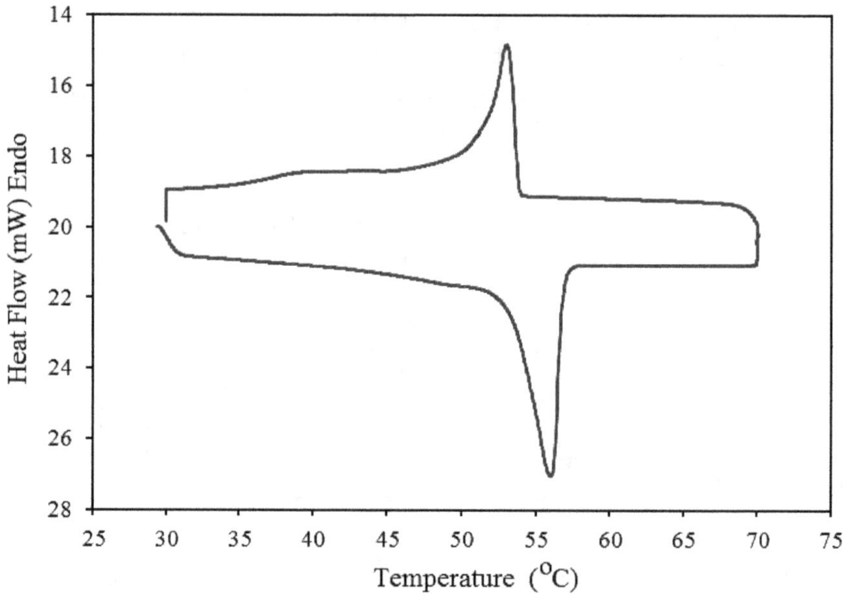

Figure 6.17 DSC thermograms of attapulgite/(PA-SA) composite as FS-CPCM.

Long-term Cycling Stability of the Developed FS-CPCM

The long-term thermal cycling stability is one of discriminating criteria for PCMs for their practical TES applications in buildings. Thus, a newly developed composite PCM should be stable in terms of chemical and LHTES properties after extended number of heating/cooling cycles [10,12,29]. The thermal cycling stability of the FS-CPCM was evaluated by subjecting the samples to accelerated melting/freezing runs. After the cycling test including 500 heating/cooling processes, the chemical structure, phase change temperatures and latent heat capacities of the FS-CPCM were compared by examining the FTIR and DSC results.

As seen from Figure 6.18, no change in the wavenumber values and shape of the main characteristic peaks of the FS-CPCM were observed after cycling period. This indicated that the long-term cycling process had no damaging effect on the chemical structure of the developed FS-CPCM.

Additionally, as seen from Table 6.3, the melting and solidification temperatures of attapulgite/(PA-SA) composite were diminished in the range of (-0.06) - (+0.01) °C, as its LHTES capacities of melting and freezing were decreased as little as 4.7% and 3.2%, respectively. Based on these results, it can be confirmed that the changes in the LHTES properties of the prepared FS-CPCM cycling were not considerable even after subjecting to repeated 500 melting-solidification cycles.

Table 6.2 Comparison of the LHTES properties of the FS-CPCM with literature [93]

Attapulgite based-FS-CPCM	Melting temperature (°C)	Latent heat of melting (J/g)	Solidification temperature (°C)	Latent heat of solidification (J/g)	Ref.
n-O/attapulgite based-FS-CPCM	24.0	101.9	29.4	101.4	[89]
S-SA/attapulgite based-FS-CPCM	55.9	69.06	-	-	[91]
M-SA/attapulgite based-FS-CPCM	59.6	68.44	-	-	[91]
Paraffin/attapulgite based-FS-CPCM	55.0	51.4	60.0	49.9	[88]
CA-PA/attapulgite based-FS-CPCM	25.09	48.21	-	-	[87]
SA-CA/attapulgite based-FS-CPCM	21.8	72.6	20.3	71.9	[92]
SA/P-attapulgite based-FS-CPCM	57.64	46.33	51.24	41.25	[92]
SA/S-attapulgite based-FS-CPCM	58.93	73.24		76.58	[92]
PA/P-attapulgite based-FS-CPCM	56.27	45.59	50.11	41.79	[92]
PA/S-attapulgite based-FS-CPCM	69.87	79.87	54.04	82.36	[92]
MA/P-attapulgite based-FS-CPCM	64.93	54.63	57.38	50.23	[92]
MA/S-attapulgite based-FS-CPCM	59.52	74.18	48.39	72.57	[92]
Attapulgite/(PA-SA) based-FS-CPCM	53.26	73.64	53.83	-72.14	This study

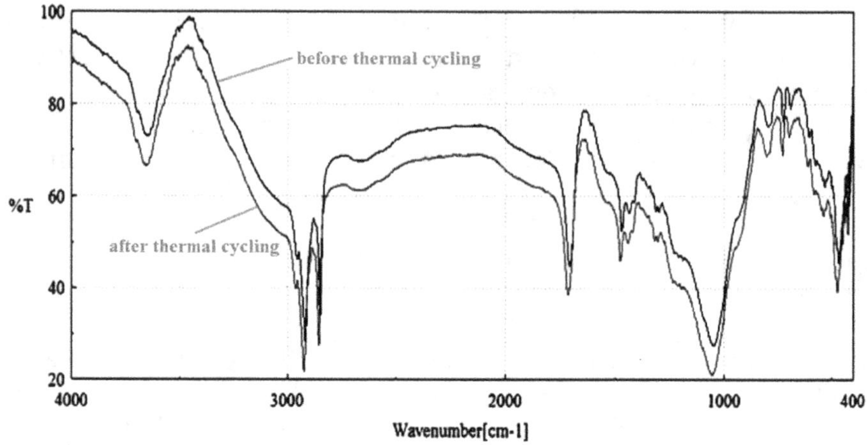

Figure 6.18 The FTIR analysis results of the attapulgite based FS-CPCM after 500 thermal cycling.

Table 6.3 The LHTES properties of the FS-CPCM after 500 thermal cycling

Melting temperature (°C)	Latent heat of melting (J/g)	Solidification temperature (°C)	Latent heat of solidification (J/g)
53.20	70.16	53.84	-69.86
The change in LHTES properties compared to before thermal cycling			
In melting temperature (°C)	In latent heat of melting (%)	In solidification temperature (°C)	In latent heat of solidification (%)
-0.06	0.01	-4.7	-3.2

6.5 Conclusions

Fatty acids as PCMs are gaining attraction for TES applications in buildings due to their efficient LHTES characteristics such as high latent heat capacity, suitable phase change temperature, good thermal stability and thermal reliability, no phase separation, slight subcooling, relatively little corrosivity, low price, etc. The eutectic mixtures of such type of PCMs allow for tailoring their phase change temperatures according to different passive solar TES targets. The integration of a eutectic mixture of two fatty acids with a lightweight traditional construction material in form-stable combination results in a composite PCM with TES ability for solar heating and cooling in building envelopes. In this respect, this chapter presents the preparation of PA-SA mixture, and its impregnation into attapulgite by vacuum technique. The specific surface area of the attapulgite was determined using BET method. The chemical and morphological properties of newly developed attapulgite/(PA-SA) composite were investigated using XRD analysis, SEM with EDX and FTIR techniques. The LHTES properties and thermal reliability of the fabricated FS-CPCM were analyzed by DSC.

The chemical and morphological results indicated that the interconnected porous structure of attapulgite acted as capillary to the fatty acid eutectics for uniform adsorption. The optimum absorption ratio of PA-SA by attapulgite was confined in the mass fraction of 40 wt% without observing any melted leakage. The eutectic mixtures were held into attapulgite by only physical attraction instead of any chemical reaction. The DSC analysis results revealed that the produced melting temperature and latent heat capacity of attapulgite/(PA-SA) composite PCM was measured as 53.26 °C and 73.64 J/g. The observed LHTES properties were observed to be suitable for solar space heating, ventilating and air conditioning purposes. The produced FS-CPCM showed good thermal and chemical reliability after 500 times repeated-thermal cycling tests. Overall, it was also concluded that the developed FS-CPCM is an effective LHTES building material to reduce energy consumption depending on the climatic conditions.

Acknowledgement

The authors would like to thank Gaziosmanpaşa University, Karadeniz Technical University and King Fahd University of Petroleum and Mineral (KFUPM) for their support. Dr. Mohammed Ouikhalfan also thanks Scientific and Technological Research Council of Turkey (TÜBİTAK) for its scholarship support under 2213-Programme.

Nomenclature

PA	Palmitic acid
SA	Stearic acid
EM	Eutectic mixture
XRD	X-ray diffraction
FTIR	Fourier transform infrared spectroscopy
DSC	Differential scanning calorimetry

SEM/EDX	Scanning electronic microscope/energy dispersive X-ray spectroscopy
PCM	Phase change material
FS-CPCM	Form-stable composite PCM
BET	Brunauer-Emmet -Teller
HVAC	Heating, cooling, ventilating and air conditioning
TES	Thermal energy storage
LHTES	Latent heat thermal energy storage
MMT	Montmorillonite
EP	Expanded perlite
FA	Fly ash

References

1. Rao, V. V., Parameshwaran, R., and Ram, V. V. (2018) PCM-mortar based construction materials for energy efficient buildings: A review on research trends. *Energy and Buildings*, **158**, 95-122.
2. Parameshwaran, R., Kalaiselvam, S., Harikrishnan, S., and Elayaperumal, A. (2012) Sustainable thermal energy storage technologies for buildings: a review. *Renewable and Sustainable Energy Reviews*, **16**, 2394-2433.
3. Aguiar, J., Cunha, S., and Kheradmand, M. (2015) Mortars with phase change materials: contribute to sustainable construction. *Key Engineering Materials*, **634**, 3-13.
4. Zhang, H., Baeyens, J., Caceres, G., Degreve, J., and Lv, Y. (2016) Thermal energy storage:Recent developments and practical aspects. *Progress in Energy and Combustion Science*, **53**, 1-40.
5. Sari, A., and Karaipekli, A. (2009) Preparation, thermal properties and thermal reliability of palmitic acid/expanded graphite composite as form-stable PCM for thermal energy storage. *Solar Energy Materials and Solar Cells*, **93**, 571-576.
6. Sari, A., and Kaygusuz, K. (2002) Thermal performance of a eutectic mixture of lauric and stearic acids as PCM encapsulated in the annulus of two concentric pipes. *Solar Energy*, **72**, 493-504.
7. Sharma, A., Tyagi, V. V., Chen, C. R., and Buddhi, D. (2009) Review on thermal energy storage with phase change materials and applications. *Renewable and Sustainable Energy Reviews*, **13**, 318-345.
8. Kuznik, F., David, D., Johannes, K., and Roux, J. J. (2011) A review on phase change materials integrated in building walls. *Renewable and Sustainable Energy Reviews*, **15**, 379-391.
9. Tyagi, V. V., and Buddhi, D. (2007) PCM thermal storage in buildings: A state of art. *Renewable and Sustainable Energy Reviews*,

11, 1146-1166.

10. Farid, M. M., Khudhair, A. M., Razack, S. A. K., and Al-Hallaj, S. (2004) A review on phase change energy storage: Materials and applications. *Energy Conversion and Management*, **45**, 1597-1615.
11. Akeiber, H., Nejat, P., Majid, M. Z. A., Wahid, M. A., Jomehzadeh, F., Zeynali, F. I., Calautit, J. K., Hughes, B. R., and Zaki, S. A. (2016) A review on phase change material (PCM) for sustainable passive cooling in building envelopes. *Renewable and Sustainable Energy Reviews*, **60**, 1470-1497.
12. Nazir, H., Batool, M., Osorio, F. J. B., Isaza-Ruiz, M., Xu, X., Vignarooban, K., Phelan, P., Inamuddin, and Kannan A. M. (2019) Recent developments in phase change materials for energy storage applications: A review. *International Journal of Heat and Mass Transfer*, **129**, 491-523.
13. Jiang, D. H., Li, A. G., Shi, F. E., and Ren, R. S. (2010) Mineral sepiolite energy-saving residential materials. *Advanced Materials Research*, **178**, 185-190.
14. Parameshwaran, R., Harikrishnan, S., and Kalaiselvam, S. (2010) Energy efficient PCM-based variable air volume air conditioning system for modern buildings. *Energy and Buildings*, **42**, 1353-1360.
15. Cao, L., Su, D., Tang, Y., Fang, G., and Tang, F. (2015) Properties evaluation and applications of thermal energy storage materials in buildings. *Renewable and Sustainable Energy Reviews*, **48**, 500-522.
16. Khadiran, T., Hussein, M.Z., Zainal, Z., and Rusli, R. (2016) Advanced energy storage materials for building applications and their thermal performance characterization: a review. *Renewable and Sustainable Energy Reviews*, **57**, 916-928.
17. Soares, N., Costa, J. J., Gaspar, A. R., and Santos, P. (2013) Review of passive PCM latent heat thermal energy storage systems towards buildings' energy efficiency. *Energy and Buildings*, **59**, 82-103.
18. Memon, S. A. (2014) Phase change materials integrated in building walls: a state of the art review. *Renewable and Sustainable Energy Reviews*, **31**, 870-906.
19. Ling, T. C., and Poon, C. S. (2013) Use of phase change materials for thermal energy storage in concrete: An overview. *Construction and Building Materials*, **46**, 55-62.
20. Lee, T. (1998) *Latent and Sensible Heat Storage in Concrete Blocks*, Masters Thesis, Concordia University, Canada.
21. Hadjieva, M., Stoykov, R., and Filipova, T. (2000) Composite salt-hydrate concrete system for building energy storage. *Renewable Energy*, **19**, 111-115.
22. Zhang, D., Li, Z., Zhou, J., and Wu, K. (2004) Development of thermal energy storage concrete. *Cement and Concrete Research*, **34**,

927-934.
23. Bentz, D., and Turpin, R. (2007) Potential applications of phase change materials in concrete technology. *Cement and Concrete Composites*, **29**, 527-532.
24. Hunger, M., Entrop, A. G., Mandilaras, I., Brouwers, H. J. H., and Founti, M. (2009) The behavior of self-compacting concrete containing micro-encapsulated phase change materials. *Cement and Concrete Composites*, **31**, 731-743.
25. Hawes, D. W. (1991) *Latent Heat Storage in Concrete*, PhD Thesis, Concordia University, Canada.
26. Xie, N., Luo, J., Li, Z., Huang, Z., Gao, X., Fang, Y., and Zhang, Z. (2019) Salt hydrate/expanded vermiculite composite as a form-stable phase change material for building energy storage. *Solar Energy Materials and Solar Cells*, **189**, 33-42.
27. Sun, D. (2010) *Study on Preparation and Properties of Paraffin Expanded Perlite Phase Change Material*, Masters Thesis, Dalian University of Technology, China [in Chinese].
28. Memon, S., Liao, W., Yang, S., Cui, H., and Shah, S. (2015) Development of composite PCMs by incorporation of paraffin into various building materials. *Materials*, **8**(2), 499-518.
29. Sarı, A., and Biçer, A. (2012) Thermal energy storage properties and thermal reliability of some fatty acid esters/building material composites as novel form-stable PCMs. *Solar Energy Materials and Solar Cells*, **101**, 114-122.
30. Sarı, A., and Biçer, A. (2012) Preparation and thermal energy storage properties of building material-based composites as novel form-stable PCMs. *Energy and Buildings*, **51**, 73-83.
31. Deng, J., Li, W., and Jiang, D. (2011) Study on binary fatty acids/sepiolite composite phase change material. *Advanced Materials Research*, **374-377**, 807-810.
32. Li, C., Fu, L., Ouyang, J., and Yang, H. (2013) Enhanced performance and interfacial investigation of mineral-based composite phase change materials for thermal energy storage. *Scientific Reports*, **3**, 1908.
33. Sun, D., and Wang, L. (2015) Utilization of paraffin/expanded perlite materials to improve mechanical and thermal properties of cement mortar. *Construction and Building Materials*, **101**, 791-796.
34. Shin, D., and Banerjee, D. (2015) Enhanced thermal properties of SiO_2nanocomposite for solar thermal energy storage applications. *International Journal of Heat and Mass Transfer*, **84**, 898-902.
35. Song, S., Dong, L., Chen, S., Xie, H., and Xiong, C. (2014) Stearic-capric acid eutectic/activated attapulgite composite as form-stable phase change material for thermal energy storage. *Energy Conversion and Management*, **81**, 306-311.
36. Yang, Y., Zhao, D., Mingwei, T., Wei, Z., and Han, W. (2014) Prepara-

tion and thermal performance characteristics of cetyl alcohol-decylic acid/fly ash shape-stabilized phase change materials. *New Chemical Materials*, 95-98 [in Chinese].

37. Sun, Z., Kong, W., Zheng, S., and Frost, R. L. (2013) Study on preparation and thermal energy storage properties of binary paraffin blends/opal shape-stabilized phase change materials. *Solar Energy Materials and Solar Cells*, **117**, 400-407.
38. Sarı, A., Alkan, C., Biçer, A., and Bilgin, C. (2014) Latent heat energy storage characteristics of building composites of bentonite clay and pumice sand with different organic PCMs. *International Journal of Energy Research*, **38**, 1478-1491.
39. Li, X., Sanjayan, J. G., and Wilson, J. L. (2014) Fabrication and stability of form-stable diatomite/paraffin phase change material composites. *Energy and Buildings*, **76**, 284-294.
40. Qian, T., Li, J., Min, X., Deng, Y., Guan, W., and Ning, L. (2015) Diatomite: a promising natural candidate as carrier material for low, middle and high temperature phase change material. *Energy Conversion and Management*, **98**, 34-45.
41. Zhang, N., Yuan, Y., Yuan, Y., Li, T., and Cao, X. (2014) Lauric-palmitic-stearic acid/expanded perlite composite as form-stable phase change material: preparation and thermal properties. *Energy and Buildings*, **82**, 505-511.
42. Jiao, C., Ji, B., and Fang, D. (2012) Preparation and properties of lauric acid-stearic acid/expanded perlite composite as phase change materials for thermal energy storage. *Materials Letters*, **67**(1), 352-354.
43. Liu, S., and Yang, H. (2015) Composite of coal-series kaolinite and capric-lauric acid as form stable phase-change material. *Energy Technology*, **3**(1), 77-83.
44. Qiu, Y., Wu, Q., Li, S., and Zhang, C. (2013) Preparation and thermal properties of binary organic/kaolin composites as shape-stabilized phase change material for thermal energy storage. *Journal of Materials Science and Engineering*, 268-272 [in Chinese].
45. Song, S., Dong, L., Zhang, Y., Chen, S., Li, Q., Guo, Y., Deng, S., Si, S., and Xiong, C. (2014) Lauric acid/intercalated kaolinite as form-stable phase change material for thermal energy storage. *Energy*, **76**, 385-389.
46. Sarı, A. (2015) Fabrication and thermal characterization of kaolin-based composite phase change materials for latent heat storage in buildings. *Energy and Buildings*, **96**, 193-200.
47. Li, C., Fu, L., Ouyang, J., Tang, A., and Yang, H. (2015) Kaolinite stabilized paraffin composite phase change materials for thermal energy storage. *Applied Clay Science*, **115**, 212-220.
48. Lei, L., and Song, B. (2015) The preparation and characterization of diatomite supported composite phase change material. *Henan Sci-*

ence & Technology, 131-134 [in Chinese].

49. Fu, X., Liu, Z., Xiao, Y., Wang, J., and Lei, J. (2015) Preparation and properties of lauric acid/ diatomite composites as novel form-stable phase change materials for thermal energy storage. *Energy and Buildings*, **104**, 244-249.
50. Jeong, S. G., Jeon, J., Lee, J. H., and Kim, S. (2013) Optimal preparation of PCM/diatomite composites for enhancing thermal properties. *International Journal of Heat and Mass Transfer*, **62**, 711-717.
51. Tang, F., Su, D., Tang, Y., and Fang, G. (2015) Synthesis and thermal properties of fatty acid eutectics and diatomite composites as shape-stabilized phase change materials with enhanced thermal conductivity. *Solar Energy Materials and Solar Cells*, **141**, 218-224.
52. Karaman, S., Karaipekli, A., Sarı, A., and Biçer, A. (2011) Polyethylene glycol (PEG)/diatomite composite as a novel form-stable phase change material for thermal energy storage. *Solar Energy Materials and Solar Cells*, **95**, 1647-1653.
53. Jiang, D., Zhao, Y., Shi, F., and Ren, R. (2010) Preparation and properties of lauric acid/sepiolite composite phase change material. *Guangdong Chemical Industry*, 37-38 [in Chinese].
54. Jiang, D., Zhou, Y., and Luo, K. (2008) Study on sepiolite composite phase-transformation building material for energy storage. *Non-Metallic Mines*, 4-5 [in Chinese].
55. Jia, Y. (2011) Reach on the organics/sepiolite fibers composite phase change heat-storage materials. Master: Hebei University of Technology, China [in Chinese].
56. Wu, K., Wang, H., Li, W., Zuo, P., and Liu, X. (2002) Determination of properties neopentyl glycol/sepiolite composites for heat storage. *Journal of Hebei Normal University*, 169-71 [in Chinese].
57. Guo, Z., Jia, Y., and Wang, X. (2010) Research on the Capric Acid/Paraffin and Sepiolite Combination as Phase Change Energy Storage Materials. *The 7th National Conference on Functional Materials and Applications*, pp. 1927-1931.
58. Wang, Y., Zheng, H., Feng, H.X., and Zhang, D.Y. (2012) Effect of preparation methods on the structure and thermal properties of stearic acid/activated montmorillonite phase change materials. *Energy and Buildings*, **47**, 467-473.
59. Chen, C., Liu, X., Liu, W., and Ma, M. (2014) A comparative study of myristic acid/bentonite and myristic acid/eudragit L100 form stable phase change materials for thermal energy storage. *Solar Energy Materials and Solar Cells*, **127**, 14-20.
60. Othman, R., Vladisavljevic, G.T., Thomas, N.L., and Nagy, Z.K. (2016) Fabrication of composite poly(d.l-lactide)/montmorillonite nanoparticles for controlled delivery of acetaminophen by solvent-displacement method using glass capillary microfluidics. *Colloids and Surfaces B: Biointerfaces*, **141**, 187-195.

61. Guo, Y., Dong, L., Zhang, Z., Fang, X., and Fang, Y. (2010) Performance research on bentonite based composite phase change thermal energy storage material with different octadecane mass fraction. *Non-Metallic Mines*, 42-44 [in Chinese].
62. Li, Z. (2005) *Study on the Preparation and Characterization of Fatty Acids/Montmorillonite Composite Materials for Phase Change Energy Storage*, Masters Thesis, Hefei University of Technology, China [in Chinese].
63. Shen, Y., and Xu, L. (2007) Preparation for neopentyl glycol/bentonite composite heat storage material. *Materials Science and Technology*, 350-353 [in Chinese].
64. Sun, D., Wang, L., and Li, C. (2013) Preparation and thermal properties of paraffin/expanded perlite composite as form-stable phase change material. *Materials Letters*, **108**, 247-250.
65. Wen, R., Zhang, X., Huang, Y., Yin, Z., Huang, Z., Fang, M., Liu Y., and Wu, X. (2017) Preparation and properties of fatty acid eutectics/expanded perlite and expanded vermiculite shape-stabilized materials for thermal energy storage in buildings. *Energy and Buildings*, **139**, 197-204.
66. Jiesheng, L., Yuanyuan, Y., and Xiang, H. (2016) Research on the preparation and properties of lauric acid/perlite phase change materials. *Energy and Buildings*, **110**, 108-111.
67. Lu, Z., Zhang, J., Sun, G., Xu, B., Li, Z., and Gong, C. (2015) Effects of the form-stable expanded perlite/paraffin composite on cement manufactured by extrusion technique. *Energy*, **82**, 43-53.
68. Wang, Q., Zhang, J., Da, J., Zhao, H., and Ran, K. (2012) Study on the thermal properties of paraffin/expansion perlite composite phase change mortar. *Advanced Materials Research*, 1274-1277.
69. Shang, J., Zhao, P., and Liu, J. (2007) Study on expanded perlite/organic acid composite phase change materials. *Journal of Xi'an University of Architecture and Technology*, 680-683 [in Chinese].
70. Sarı, A., and Karaipekli, A. (2008) Preparation, thermal properties and thermal reliability of capric acid/expanded perlite composite for thermal energy storage. *Materials Chemistry and Physics*, **109**, 459-464.
71. Wang, Z. (2013) *Preparation and Properties of Fatty Acids Supported by Expanded Perlite Composite Phase Change Materials for Application in Diving Suit*, Masters Thesis, East China University of Science and Technology, China [in Chinese].
72. Kant, K., Shukla, A., and Sharma, A. (2016) Ternary mixture of fatty acids as phase change materials for thermal energy storage applications. *Energy Reports*, **2**, 274-279.
73. Yu, Y. (2011) *Preparation and Application of Phase Change Energy Storage Composite Materials Based on Expanded Perlite*, Masters

Thesis, Xinyang Normal University, China [in Chinese].

74. Lu, Z., Xu, B., Zhang, J., Zhu, Y., Sun, G., and Li, Z. (2014) Preparation and characterization of expanded perlite/paraffin composite as form-stable phase change material. *Solar Energy*, **108**, 460-466.
75. Karaipekli, A., Sarı, A., and Kaygusuz, K. (2009) Thermal characteristics of paraffin/expanded perlite composite for latent heat thermal energy storage. *Energy Sources Part A: Recovery, Utilization, and Environmental Effects*, **31**, 814-823.
76. Wang, W., Kang, H., Yang, X., Fang, Y., and Ding, J. (2007) Study on polyethylene glycol /silicon dioxide composite phase change thermal energy storage material. *Journal of Functional Materials*, 1652-1657 [in Chinese].
77. Wang, Y., Xia, T. D., Zheng, H., and Feng, H. X. (2011) Stearic acid/silica fume composite as form stable phase change material for thermal energy storage. *Energy and Buildings*, **43**(9), 2365-2370.
78. Xu, B., Ma, H., Lu, Z., and Li, Z. (2015) Paraffin/expanded vermiculite composite phase change material as aggregate for developing lightweight thermal energy storage cement based composites. *Applied Energy*, **160**, 358-367.
79. Chung, O., Jeong, S., and Kim, S. (2015) Preparation of energy efficient paraffinic PCMs/expanded vermiculite and perlite composites for energy saving in buildings. *Solar Energy Materials and Solar Cells*, **137**, 107-112.
80. Karaipekli, A., and Sarı, A. (2010) Preparation, thermal properties and thermal reliability of eutectic mixtures of fatty acids/expanded vermiculite as novel form-stable composites for energy storage. *Journal of Industrial and Engineering Chemistry*, **16** (5), 767-773.
81. Karaipekli, A., and Sarı, A. (2009) Capric-myristic acid/vermiculite composite as form-stable phase change material for thermal energy storage. *Solar Energy*, **83** (3), 323-332.
82. Zhang, N., Yuan, Y., Li, T., Cao, X., and Yang, X. (2015) Study on thermal property of lauric-palmitic-stearic acid/vermiculite composite as form-stable phase change material for energy storage. *Advances in Mechanical Engineering*, **7**(9).
83. Jiang, D., Shi, F., Ren, R., and Zhao, Y. (2010) Experimental research on coal ash composite for energy storage. *Energy Environmental Protection*, **33**, 28-29 [in Chinese].
84. Zhang, D., Tian, S., and Xiao, D. (2007) Experimental study on the phase change behavior of phase change material confined in pores. *Solar Energy*, **81** (5), 653-660.
85. Hua, J., Wang, J., Cheng, H., Li, F., and Xue, C. (2013) Availability of preparation process of high temperature fly ash based phase change composite. *Hot Working Technology*, **101**, 96-98 [in Chinese].

86. Wang, J. (2013) *Study on Preparation and Properties of High Temperature Fly Ash Based Phase Change Heat Storage Composite Materials*, Masters Thesis, Xi'an University of Architecture and Technology, China [in Chinese].
87. Li, M., Wu, Z., and Kao, H. (2011) Study on preparation, structure and thermal energy storage property of capric-palmitic acid/attapulgite composite phase change materials. *Applied Energy*, **88** (9), 3125-3132.
88. Shi, T., Sun, W., and Wang, Q. (2009) Study on preparation and thermal performance tests of attapulgite-based phase change composites for energy storage. *Acta Materiae Compositae Sinica*, 143-147 [in Chinese].
89. Chen, P., Zhang, H., Sun, L., Cao, Z., Xu, F., Chen, J.,and Wu, T. (2014) Preparation and thermal properties of N-octadecane/attapulgite composite phase change materials. *Materials Reviews*, 62-66 [in Chinese].
90. Lv, P., Ding, M., Liu, C., and Rao, Z., (2019) Experimental investigation on thermal properties and thermal performance enhancement of octadecanol/expanded perlite form stable phase change materials for efficient thermal energy storage. *Renewable Energy*, **131**, 911-922.
91. Wang, Y., Zheng, H., Feng, H., and Xia, T. (2011) Preparation and characterization of stearic acid/activate attapulgite composite phase change materials. *Modern Chemical Industry*, 43-46 [in Chinese].
92. Liang, W., Chen, P., Sun, H., Zhu, Z., and Li, A. (2014) Innovative spongy attapulgite loaded with n-carboxylic acids as composite phase change materials for thermal energy storage. *RSC Advances*, **4**, 38535-38541.
93. Lv, P., Liu, C., and Rao, Z. (2017) Review on clay mineral-based form-stable phase change materials: Preparation, characterization and applications. *Renewable and Sustainable Energy Reviews*, **68**, 707-726.
94. Hussain, S. I., Dinesh, R., Roseline, A. A., Dhivya, S., and Kalaiselvam, S. (2017) Enhanced thermal performance and study the influence of sub cooling on activated carbon dispersed eutectic PCM for cold storage applications. *Energy and Buildings*, **143**, 17-24.

Index

A

B

C

D

E

Index

F

G

H

I

L

M

N

O

P

R

S

T

V

W

www.ingramcontent.com/pod-product-compliance
Lightning Source LLC
LaVergne TN
LVHW011521200926
848484LV00001B/93

* 9 7 8 1 9 2 5 8 2 3 3 9 4 *